THE QUEST FOR ALIEN PLANETS

EXPLORING WORLDS OUTSIDE THE SOLAR SYSTEM

THE QUEST FOR ALIEN PLANETS

EXPLORING WORLDS OUTSIDE THE SOLAR SYSTEM

PAUL HALPERN

PLENUM TRADE · NEW YORK AND LONDON

Library of Congress Cataloging-in-Publication Data

Halpern, Paul, 1961-
    The quest for alien planets : exploring worlds outside the solar
  system / Paul Halpern.
       p.   cm.
    Includes bibliographical references and index.
    ISBN 0-306-45623-0
    1. Life on other planets. 2. Human-alien encounters.   I. Title.
  QB54.H285   1997
  523--dc21   97-22445
                                                                   CIP

ISBN 0-306-45623-0

© 1997 Paul Halpern
Plenum Press is a Division of Plenum Publishing Corporation
233 Spring Street, New York, N.Y. 10013-1578
http://www.plenum.com

10 9 8 7 6 5 4 3 2 1

Printed in the United States of America

For Eli, the explorer

There must be an infinite number of suns with planets with life on them.

> —Giordano Bruno, Italian monk,
> burned at the stake in 1600

Preface
WHO ARE THE ALIENS?

Bug-eyed monsters with slimy, green skin and long, sharp talons, stomping muddy tracks through eerie marshes. Emotionless beings with raised eyebrows and pointed ears. Silver-coated androids running amok, blasting innocent Earthlings with powerful laser beams. Amorphous blobs that penetrate even the sturdiest of barriers. Doll-like, impish creatures with wrinkled skin, wanting only to "phone home." And telepathic energy fields that long for contact with corporeal races such as our own, silently hoping to become our friends.

These are some of the diverse images we have of extraterrestrials. They derive mainly from popular sources: speculative novels, plays, television programs, movies, and the like. For many years, a booming, Hollywood-based science fiction film industry has generated the costumes, props, and special effects needed to produce the illusion of life on other planets. Writers have spent their careers churning out page after page, describing imaginary beings from other worlds. Fictitious "alien" imagery is omnipresent—from *Star Wars* dolls in toy stores, to life-size cardboard cut-outs of *Star Trek* characters in book shops. And, years after its release, *E.T. the*

Extraterrestrial is still one of the most financially successful and critically acclaimed children's films of all time.

Still, the public wants to learn more and more about
mysterious "space creatures." A recently opened exhibit
about aliens at the family vacation mecca, *Walt Disney
World*, has already become one of the park's most popular
attractions. Called the ExtraTERRORestrial Alien Encounter, it invites visitors to witness a life form from
a distant planet. Thrill-seeking tourists sometimes line
up for an hour or more for this strange, otherworldly
experience.

The ride begins with the participants being whisked
into a foreboding circular chamber where they are strapped
into special seats and prepared for the experience of interstellar contact. There, they are instructed about a new
scheme for the instantaneous conveyance of objects from
planet to planet. A subsequent demonstration of interplanetary transport goes awry, and a hideous alien is
brought to the chamber. Its horrid form suddenly appears
within a glass containment vessel, right in front of startled
onlookers. Soon—as any horror buff would expect—it
escapes. As the audience gasps from sheer terror (or
chuckles, as the case may be), it breaks out of its containment vessel. The room immediately becomes pitch black,
adding to the suspense. State-of-the-art auditory, visual,
and tactile effects enable frightened participants to feel
they are being attacked by this hostile being. After the
encounter is over audience members often leave the exhibit visibly shaken. (Children have been known to
scream and beg their parents to be let out of the room well
before the show is over.) Still, the crowds keep streaming
into this popular attraction—located in the building that
formerly housed another fictitious space show, the venerable Mission to Mars ride.

Why does the public hunger to confront aliens? Why
are media depictions of extraterrestrial life so captivating

as shown by the popularity of such television shows and films as *Sightings, Alien Nation, A.L.F., The X-Files, Third Rock from the Sun, Aliens, Independence Day, Mars Attacks,* and several *Star Trek* series? Perhaps it is because—to borrow an expression from *Star Trek* that has grown into a cliché—space is the final frontier. We no longer send explorers halfway around the globe to seek out hidden civilizations and unknown peoples. Adventurers and anthropologists have raked the Earth over and over for signs of a yet undiscovered culture. Maybe there are still a few primitive tribes to be found, hidden from advanced society in remote jungle regions. In all probability, the cultural map of our world is essentially complete. The days of a Jonathan Swift writing about bizarre peoples (Lilliputians, etc.) living in unexplored parts of the globe are over. We need to set out into space to encounter beings and cultures that are radically different from what is already known.

In the centuries to come, as terrestrial culture becomes increasingly homogeneous, this longing for extraterrestrial contact will likely grow even stronger. With McDonald's restaurants in the Middle East and Coke machines throughout the Pacific Islands, trips to other planets (or imagined ones at least) may enable us to experience the exotic "lifestyles" radically dissimilar from our own.

When we finally do establish contact with alien life forms, assuming they exist, we will no doubt be amazed. Our current mental pictures of what extraterrestrials may be like stem mainly from depictions in fiction. Forming the stuff of cheap science fiction novels, television series schlock, and Hollywood B-grade movies, they are based purely on conjecture and longing. The reason for this lack of substance is simple. Authors' preconceptions are shaped necessarily by their own experiences. Even the most imaginative of writers cannot escape the fact that they were born on Earth and will likely never leave it.

Our utter lack of knowledge about alien life forms means if contact were to occur, we would be even less prepared than many of our ancestors were when they initially encountered societies more advanced than their own. In cases when less advanced peoples (of the Americas, Africa, and Oceania, for example) first came into contact with technologically superior nations (Europeans, for instance), they experienced strong culture shock. Often they were so confused and overwhelmed by their new circumstances they became ripe for exploitation—or extinction.

Imagine how the indigenous groups of remote Borneo, used to traveling exclusively on foot, felt when they saw their first airplane, or for natives of the Amazon basin, their first cameras. Almost immediately, their views of the world—of everything they knew and cherished—became disrupted beyond repair. It would have been virtually impossible for them to forget the experience and start over. Once their illusions were shattered, there was no turning back.

Considering the tremendous impact that the coming of Europeans had on aborigines in isolated parts of the globe, it is hard to fathom the far greater effects on both species that an interplanetary encounter would represent. This would especially be the case if the alien race possessed intelligence superior to ours. Should the other species be more advanced than we (and we need not assume this), then their existence would present a peril. We might only hope that the human race would survive such a strong challenge to its identity.

Am I suggesting that humankind would be "zapped" out of existence by alien weapons? Hardly. In my view, the far greater threat would be to our pride. We are used to being iconoclasts, to fending for ourselves in our own minuscule part of space. We think of ourselves as special—the lone bearers of the gift of conscious intelligent thought. Moreover, we traditionally consider ourselves the reason why the universe itself was created. The mere existence of extraterrestrials would shatter irrevocably this human-centered view of the cosmos.

Even in our space travel fantasies, we often imagine that our race will someday be anointed kings and queens of the universe, or at least of our galaxy. According to these geocentric visions, we would merely have to leave Earth, encounter aliens and flaunt our superior qualities to them, in a kind of cosmic debutantes' ball, and they would bow down to us.

Consider, for instance, the various *Star Trek* science fiction series. Though in these programs, many strange alien races are depicted, the bulk of those shown look and act very much like humans. Moreover, it is the prime goal of many of the fictional aliens presented to study, emulate, and even become humans. And it is most telling that humans are portrayed as dominating a future Galactic Federation, which is headquartered in San Francisco.

In truth, how likely is it that extraterrestrials would come to think of San Francisco, of all places, as their capital? Not very. It's a spectacular place, no doubt, with stunning vistas. But it's hardly central to places like Betelgeuse or Sirius. (It's not even central to Kansas City or Milwaukee, let alone the far reaches of the galaxy.)

In Isaac Asimov's well-respected *Foundation* series, on the other hand, the reigns of the galaxy are controlled by an empire centered on a distant planet called Trantor. Yet not even a visionary writer like Asimov could resist anthropomorphic temptations. The series gradually reveals that all of the galaxy's peoples can trace their origins back to a single planet—Earth!

Yet it's hard to fault science fiction for its pervasive geocentrism in making the assumption that Earth is the center of the universe. One might argue that one of the main functions of science fiction is to provide a thinly disguised version of earthly events. Often, it furnishes a useful forum for a "what if" look at our own history and culture. Thus we must consider its admonitions to be parables, for the most part, rather than prognostications.

If we were ever on the verge of meeting the real extraterrestrial denizens of our galaxy, we might not expect as easy an encounter as contemporary science fiction frequently suggests. Rather, we would have to brace ourselves for an experience for which there could be little true preparation.

If contact with aliens were to start through radio transmissions, our initial knowledge would be limited to what could be gleaned through broadcasts. Forming a picture of them in this manner would almost certainly prove challenging. Quite likely, in the beginning stages of contact, language barriers would make it difficult for us to obtain much information about them. Having no common ground to base our mutual communications would force us to struggle hard to fathom their messages, as well as to make ourselves understood.

All human languages, no matter how obscure, are grounded in certain commonalities. We each have a mouth, nose, two eyes, and two ears, as well as a mother and father. Each of us eats, drinks, and sleeps. Therefore, every language on Earth must have words or expressions for these body parts, relationships, and activities. When someone learns a new tongue, he or she might reasonably expect an answer to the question: "What is the word for this or that?"

Furthermore, human beings communicate with each other through standard, predictable mechanisms. In verbal and gesticulatory communication, we possess a fixed set of sounds and signs, from which each utterance must be fashioned. Moreover, a language mechanism in the brain shapes the types of grammar we can form. In written and transmitted language, we employ alphabets and/or pictographs that have similarity of origin.

Therefore, if anthropologists wish to translate a newly discovered language, they have the means at their disposal. They first look for similarities to known languages—in usage, grammar, and symbolic notation. Then, through these parallels, they form a "rosetta stone" connecting the unknown language with more familiar idioms.

With extraterrestrials, there could be no such assumptions of commonality. Although there would be some points of reference—the nature of the stars, the chemical elements, mathematics, etc.—there would be no shared experience to which we might allude. They could not draw from humanity's considerable lexicon, and we could not borrow from theirs.

Even if the language barrier were surmounted, we might find ourselves facing obstacles far more formidable. What if their entire way of thinking and being is beyond our comprehension and ours beyond theirs? Language is certainly no guarantee of true understanding. Even now on our own planet, where translators rapidly interpret the communications between nations, misunderstandings abound. And sometimes, sadly, these lead to war.

Yet for all the clashes between world nations, the situation between terrestrials and aliens would be far more volatile. Earth-dwellers, be they Africans, Europeans, Asians, or Americans, have many more commonalities than differences. As members of the same species, we harbor similar needs and desires. However, extraterrestrials, if they happened to be intelligent, would presumably possess radically dissimilar attributes. These severe differences could lead to profound difficulties, perhaps mutual hostility. Maybe, for example, the intonation patterns of human voices would horribly grate on certain alien races. Or the eating habits of some extraterrestrials might disgust and offend us, leading to attitudes, on our part, of extreme prejudice.

Let's be optimistic, however, for the purposes of this

discussion. Assuming that these potential rifts could be somehow bridged—through patience and cooperation—our society would likely be amply rewarded by contact with an extraterrestrial civilization. In the ideal case, we would soon become privileged to share in the knowledge and wisdom of a vastly different race from our own. Through comparing ourselves with the other species, we would gain insight into what it means to be intelligent.

Yet even in the ideal case of benevolent, mutually enriching contact, there would still be grave dangers for the human race. Suppose the species we encountered happened to be vastly superior to us. We might be so humbled by their greater mental and physical capacities we'd lose our resolve. Their technology might be so superior to ours we might feel like children in comparison. Their understanding of the cosmos might be so comprehensive we would lose our sense of self-reliance and become dependent on them for all scientific information. In the extreme case, we might well decide that human scholarly endeavors would be pointless in the face of more comprehensive sources of knowledge. Our desire to explore the universe might well be quashed. In short, if the masquerade of human mastery of the universe were to end, there is danger that the resulting loss of pride could wholly rob our species of its sense of destiny and purpose.

To return again briefly to the realm of fiction, consider Arthur C. Clarke's epic novel *Childhood's End*. In this dark tale, a group of alien visitors, called the Overlords, comes to Earth and cures all of its social woes. In a matter of decades, they eliminate the world's crime, warfare, and poverty, rebuild its cities, and institute world government. As they smother the human race with kindness and dazzle it with new technologies, it grows increasingly dependent on them. Ultimately, humankind becomes so spoiled by this treatment that it loses its special sparks of initiative and creativity. Science, art, and other forms of independent human expression all fall by the wayside.

In a poignant passage, Clarke describes how this intellectual decline takes place:

> There were plenty of technologists, but few original workers extending the frontiers of human knowledge. Curiosity remained, and the leisure to indulge in it, but the heart had been taken out of fundamental scientific research. It seemed futile to spend a lifetime searching for secrets that the Overlords had probably uncovered ages before ...
>
> The end of strife and conflict of all kinds has also meant the virtual end of creative art. There were myriads of performers, amateur and professional, yet there had been no really outstanding new works of literature, music, painting, or sculpture for a generation. The world was still living on the glories of a past that could never return.[1]

In summary, our biggest hope (however unrealistic) for an encounter with intelligent beings from another planet is that it would solve the outstanding problems facing the human race. On the other hand, our greatest fear for such an encounter is that it would solve the outstanding problems facing the human race. Although humankind would welcome connections with other cultures, it also thrives on challenges. One only hopes that once the shock of finding ourselves in some manner "inferior" to another species had passed, we would swallow our pride, regroup, and strive for new, loftier goals.

All this discussion is hypothetical, of course. Contact with extraterrestrials may be centuries, even millennia, away. And it is not necessarily the case that other forms of life in the cosmos possess greater or even comparable intelligence to ours. We must even consider the possibility, however improbable, that there are no other intelligent life forms in space. In that case, our pride would be reprieved, but our loneliness augmented considerably.

After all, it has only been in recent years that astronomers have firmly established there are planets around other stars. They have found at least a dozen new worlds, circling a number of distant suns. Though none of the newly found planets seem to present conditions favorable for life, these momentous discoveries, the focus of our exploration here, have increased our expectations that habitable worlds will soon be encountered. The possibility of finding living worlds has, in turn, bolstered our hopes—and apprehensions—about coming face to face with our extraterrestrial counterparts. Who are the aliens? Where are the inhabited worlds? We may find out soon enough.

ACKNOWLEDGMENTS

This book was written in part during my sabbatical year at the Philadelphia College of Pharmacy and Science. Therefore, I thank the administration and staff of this college for enabling me to have the time and resources necessary to complete this work. In particular I would like to thank Phil Gerbino, Nancy Cunningham, Carol Weiss, Maven Myers, Charles Gibley, William Walker, Elizabeth Bressi-Stoppe, Durai Sabapathi, Jennifer Mc-Greevy, Paul Angiolillo, David Kerrick, and Bernard Brunner for their support.

During the same year this book was written, my son Eli was born. I thank him then for his nap times and play times, allowing me to get some research and writing done. Moreover, I appreciate my wife Felicia's help, support, and insight—more than anyone could hope for. I thank my parents, Stanley and Bernice Halpern, and my in-laws, Joseph and Arlene Finston, for providing critical extra assistance. I also thank family members Richard, Anita, Alan, Ken, and Esther Halpern; Lane and Jill Hurewitz; Janice and Richard Antner; Fred and Jennifer Schwartz; and Shara and Richard Evans, for their support.

Whenever I need extra insight or inspiration, I lean

on my friends. I feel fortunate to have such a fine group to count on for advice. I'm particularly grateful to Michael Erlich, Fran Sugarman, Elana Doering, Simone Zelitch, Scott Veggeberg, Marcie Glicksman, Fred Schuepfer, Pam Quick, and Kris Olson. Thanks to Alex Wolszczan, for answers to my questions about pulsar planets, and to the late Clyde Tombaugh, for insightful comments about his discovery of Pluto. Wulff Heintz was extremely kind to lead me on a private tour of Sproul Observatory, to show me Peter Van de Kamp's plates of Barnard's Star, and to relate a chronology of planet hunting at Sproul. I am also grateful to Geoffrey Marcy for his detailed account of his research and Martyn Fogg for speedy answers to my questions. Thanks to Swarthmore College, Lick Observatory, New Mexico State University, San Francisco State University, NASA, and the Carnegie Institution of Washington for photographs.

Finally, I thank Linda Regan for her helpful editorial suggestions, ideas, and support and Robert Ubell for his hard work and guidance.

CONTENTS

COMMENT ABOUT TERMINOLOGY

The reader should note that common astronomical terms such as "sun," "moon," "earth," and "solar system" sometimes appear capitalized in this book and in lower case at other times. This designation is to distinguish between descriptions of our own familiar part of space versus references to more exotic locales. When such terms refer to *our* domain, then they are capitalized. Example: "the Sun, the Earth, Mars, and seven other planets all belong to the Solar System." Otherwise they are in lower case, e.g., "the search for other earths, orbiting distant suns is led by a dedicated group of astronomers." I hope that these distinctions help to minimize confusion between the mundane and the extraordinary.

Introduction
THE LURE OF
UNKNOWN WORLDS

He, who through vast immensity can pierce,
See worlds on worlds compose one universe,
Observe how system into system runs,
What other planets circle other suns,
What varied Being peoples every star,
May tell why Heaven has made us as we are.

ALEXANDER POPE, *An Essay on Man*

Are we alone? Or are there other intelligent life forms in the universe? It has been more than 35 years since scientists first began scanning the skies for messages from other worlds. Using enormous broad-dish receivers, they have searched likely radio channels again and again for signs of extraterrestrial broadcasts. Yet, in spite of their considerable efforts, they have detected no signs of intelligent life in space.

It would be unfortunate indeed if our race were solitary. Humankind longs to communicate with other beings who share their joy of discovery, their love of life, and

1

their conscious awareness of the universe. Without a doubt, a detailed knowledge of the triumphs and tragedies of other cognizant species, assuming they are sufficiently advanced, would help us immensely as we plan our future. Our venerable cousins on other planets could potentially provide us with new medicines, technologies, and inventions—perhaps even with insight into resolving our social problems. Moreover, by comparing notes with other highly evolved forms of life, we could learn much about ourselves. We could finally be able to determine which features make us unique as a species, and which are shared among all intelligent beings. What a bitter disappointment, then, that there have been no hints, not a single telltale transmission, suggesting that extraterrestrials exist.

But we cannot give up hope so soon. Although there have yet to be signs of advanced existence beyond Earth, some scientists argue that evidence for primitive life in space has already been found. In 1996, a team of NASA researchers announced that they had discovered traces of life-associated elements embedded in a meteorite of Martian origin found in Antarctica. Their analyses indicate that the complex materials found in the rock—called polycyclic aromatic hydrocarbons (chemical substances often seen as byproducts of organic processes)—were most likely produced by living organisms billions of years ago on Mars. Moreover, images of the meteorite taken with an electron microscope depict what appear to be the outlines of tiny cells. The NASA team contends that these are the traces of fossilized Martian microorganisms.

If these findings are confirmed by further testing or new discoveries the implications would be monumental. With life known to exist on two planets (Earth and Mars), rather than just one, we might reasonably infer that life in space is more abundant than once thought. And, if life is fairly common, the possibility that it developed elsewhere in the universe into other advanced beings would seem all the more likely. The assembly line of evolution that has packaged simple amoeba into complex creatures on Earth would presumably operate on other living planets as well. Let us not jump to conclusions, however. To provide definitive proof that life once existed (or even still exists) on Mars, NASA must send special excavatory missions to it. Only a comprehensive study of the red planet's rocks and soil would settle this issue.

The Martian question aside, there is good reason to believe we will eventually encounter intelligent entities in space. There are approximately 100 billion stars in our galaxy and billions of galaxies in the observable universe. Assuming (and there is substantial reason to do so) that a solid fraction of these stars are encircled by planetary systems—and that at least some of these planets harbor advanced, sentient beings—then intelligent life abounds in the cosmos.

Essentially, to seek out these inhabited worlds, one has to choose between two different methods: scanning for alien communications and hunting for alien planets. The first strategy, the longstanding radio search for extra-terrestrial intelligence (SETI), relies on the expectation that advanced alien civilizations would seek to communicate, at some point, with radio signals. SETI enthusiasts hope that by monitoring likely channels for these broadcasts, they can detect these messages and trace them back to their sources.

Why hasn't the SETI program produced results in the decades it has been running? A few scientists, notably Frank Tipler of Tulane University, have suggested that we haven't heard from alien beings because they simply aren't out there. Tipler asserts that extraterrestrials, if they existed, would have tried to colonize us a long time ago by sending an invading fleet (of robot ships, he suggests). Any race with technological know-how, he argues, would have eventually realized the need for expansion of its territories throughout the universe, and figured out a way of achieving its goal. Because this obviously hasn't happened, Tipler concludes that the only intelligent, self-aware creatures in the universe are on Earth. Therefore, he feels that the current search for extraterrestrial intelligence is a complete waste of time.

Other researchers, such as SETI guru (and Santa Cruz astronomy professor) Frank Drake, strongly disagree with Tipler and his supporters. They feel that the present search for extraterrestrials is only in its infancy, and that it should be given more time to succeed. Drake, who participated in the earliest sky scans for space broadcasts back in the 1960s, finds Tipler's arguments absurd. He questions Tipler's fundamental assumption that extraterrestrials would necessarily want to conquer space. Why, asks Drake, must we assume that other intelligent species would be as violent and greedy as some humans? Perhaps, he argues, alien races would be content just keeping to their own planets. Maybe they would prefer to advertise their presence with peaceful broadcasts, rather than with conquering fleets.

Nathan Cohen, one of Drake's students at the time when this debate was at its peak, composed a humorous letter mocking Tipler's beliefs.

"Have you seen Frank Tipler?" Cohen asks. "There are only four billion people on this planet; surely an intelligent creature would find some direct way of making his presence known to at least a sizable fraction of the population."[1]

Cohen concludes that the obvious way of establishing Tipler's existence is to search for him, implying that this would be the natural way, as well, of establishing or ruling out the existence of extraterrestrials.

"Perhaps we haven't seen Frank Tipler because we haven't looked hard enough," he writes. "If we undertook a comprehensive and methodical search for him then we may be able to make a definitive decision on his existence."[2]

Drake agrees that the best way to find out if there are other life forms in space is to look hard for them. He feels that theoretical reasoning doesn't prove anything; it just drags you around in circles. That is why he is particularly peeved that the U.S. Congress recently cut off all funding for SETI just a year after promising it ample support. Much to Drake's chagrin, many prominent individuals share Tipler's belief that SETI isn't worth the cost. Drake feels that they are making a big mistake, one that may compromise for a long time our prospects of contacting our possible counterparts on other planets.

In spite of the SETI funding cutoff, our chances of finding inhabited worlds, appear far from bleak. In the mid-1990s, NASA launched several new programs that will lead, I believe, to the eventual discovery of life in space. These include Astronomical Studies of Extrasolar Planetary Systems (ASEPS), Exploring Neighboring Planetary Systems (ExNPS), and the Planet Finder and Planet Mapper programs. Together, these projects represent an all-out effort to find Earth-like planets in space. Our best telescopes will be coupled with the finest pieces of astronomical equipment in order to discern the telltale signs of alien worlds around distant stars.

NASA's strategy is straightforward. First, astronomers will hunt for extrasolar (outside of our Solar System) planets that have approximately the same mass as Earth. After finding these worlds, they will determine if their atmospheres possess terrestrial qualities, such as oxygen content. When the technology becomes available, probes will be sent out to examine the most promising of these candidates. And finally, in the distant future, manned missions will be launched to the worlds that appear habitable.

One might think naively that the first step of this mission—seeking planets beyond the Sun—would be completely trivial. Therefore, one might even suppose that thousands of them have been found already. After all, haven't "interplanetary voyages" been presented in hundreds of popular movies and television shows? Don't spaceship captains Jean-Luc Picard and James Kirk of the various *Star Trek* series cruise from planet to planet on each episode, as if taking casual sea voyages between tropical islands? Why else have so many fictional, otherworldly place names, such as Planet Vulcan and Forbidden Planet, passed over the years into common parlance? What could be simpler, it would seem, than to map out the planets in other systems?

On the contrary, very few planetary candidates have been found so far. In fact, the first confirmed sighting of a planet has been a surprisingly recent event. It wasn't until the early 1990s that Alexander Wolszczan, a researcher studying radio pulsars (tiny, rapidly rotating, ultra-dense stellar objects), stumbled upon the first known set of extrasolar planets.

Wolszczan is the sort of person one would call a natural born astronomer. As a child in Poland, he first became interested in the nature of space while perusing the many books on the subject in his father's library. At eight, while his classmates were thumbing through reading primers with colorful pictures, he devoured every book about the stars he could find.

Naturally, he couldn't wait to get into university and begin his formal training in the field of astronomy. Aptly enough, he attended Nicholas Copernicus University in Torun, Poland (the town in which Copernicus was born). Interestingly, Copernicus was the first to assert that Earth is not the only true world (in the sense of having material substance), and Wolszczan was the first to show that the Solar System is not the only planetary system.

In the 1970s, after completing his Ph.D. in radio-astronomy (the detection and analysis of radio waves from celestial bodies), Wolszczan decided to specialize in the study of pulsars. Pulsars were still a relatively new topic at the time, having been discovered just a few years earlier. Scientists were (and continue to be) fascinated by their highly periodic emission of radio signals. These signals are given off as they spin, similar to the way pulses of light are flashed by a rotating lighthouse beacon.

To carry out his study, Wolszczan took on a research position at Arecibo Observatory in Puerto Rico, home of a huge, fully equipped radio telescope. Far from the often harsh climate of Poland, he grew to relish Arecibo for its lush tropical jungle, splendid beaches, and friendly people. There he began a detailed program to monitor the fluctuating signals of a large group of pulsars, plotting out their times of arrival.

Normally, pulsar monitoring can be pretty tedious, akin to watching a metronome for hours on end. Under ordinary circumstances, pulsars are creatures of habit, emitting the same periodic pulse again and again. Thus, work life for Wolszczan was fairly predictable, until he decided in the late 1980s to take a look at a pulsar called PSR 1257+12. He found strong evidence that this pulsar was producing signals influenced by nearby planets. Wolszczan's experience and patience enabled him to zero in on indications that were subtle at best.

It was a small thing, really, the kind of minute abnormality astronomers call a "glitch." The untrained eye surely wouldn't have noticed it, not in a million years. This type of phenomenon required an expert to resolve—along with a hefty dose of computer power.

Wolszczan was as thorough as possible in his evaluation of these anomalous signals. He checked and double-checked his equipment. Using the supercomputers at the Cornell Theory Center, he performed a detailed analysis of his data. Gradually, he eliminated all possibilities but one. He came to the startling conclusion that the pulsar is surrounded by a planetary system. These planets exert gravitational tugs on the pulsar, causing it to wobble slightly and produce an observable glitch in its light output arrival times.

Could any of these newly found pulsar planets be the home of an alien species? Not very likely; they are bathed constantly in lethal radiation. I asked Wolszczan how he imagined these worlds. His response showed an elaborate knowledge of what the planets are probably like—and they sure don't look promising for life.

"The planets are almost certainly terrestrial in their masses. They have been most probably made of highly evolved stellar material, so they may actually be 'iron planets.' If the pulsar beam sweeps past the planets, their initial atmospheres (if any) would have been wiped out long ago. It may be that the pulsar's energy heats them up to some 600–700 K (degrees above absolute zero) and that the planets are bathed in high energy radiation and streams of relativistic particles. Isn't it a fun place to be?"[3]

Picture living creatures trying to survive on a world in which the only sunshine comes in the form of regular, low-frequency energy bursts of limited duration. It is hard to imagine life thriving in a land where lead and tin would only exist in molten states. For these reasons, scientists strongly believe that the pulsar planets found by Wolszczan are barren.

After Wolszczan's discovery, astronomers redoubled their efforts to find potentially habitable planets. Even though Wolszczan's planets appeared uninhabitable, the fact that there were worlds at all in other systems seemed to auger well for the future of interplanetary exploration. Their existence strongly suggested that planets would be found near living stars as well. And presumably some of these orbiting bodies would resemble Earth in size and composition, perhaps even harboring intelligent life. It is not hard to understand why most researchers consider Wolszczan's discovery a milestone in the history of our search for habitable places in the cosmos.

Indeed, recently astronomers have been absolutely delighted to report the finding of at least nine new extrasolar planets. These additional worlds are remarkable because they circle living stars resembling our own Sun, rather than orbiting extinguished orbs. Thus, unlike Wolszczan's worlds, each of these newly found planets possesses at least one of the necessary conditions for life (as we presume it to be): a shining star nearby.

The recent spate of discoveries of star-circling planets began with the October 1995 announcement by Geneva Observatory astronomer Michel Mayor and his graduate student Didier Queloz of the sighting of a world near the star 51 Pegasi. Painstaking research went into the spotting of this planet. The Swiss team took high-precision measurements of its velocity for 18 months, and released their findings only after their statistics looked especially promising. After they announced their results at a workshop in Cambridge, a group at Lick Observatory immediately set out to provide verification. Within weeks, the second group confirmed that an extrasolar planet had indeed been found around the star in question.

The Lick Observatory team, led by Geoffrey Marcy of San Francisco State University and Paul Butler of the University of California at Berkeley, was hardly new to the science of planet hunting. This veteran group of astronomers began their search for extrasolar planets in 1987. Undoubtedly, it was disappointing for them not to be the first to find planets outside the Solar System. Not to be outdone by their Swiss rivals, however, soon after the planet around 51 Pegasi was discovered, the Lick team announced that they had spotted two more extrasolar worlds.

The planets found by Marcy, Butler, and their colleagues orbit stars situated in two familiar constellations. The first world discovered revolves around the star 47 Ursae Majoris, located 35 light years away in the Great Bear constellation. (One light year is about six trillion miles.) The second is a similar distance from Earth, but near the star 70 Virginis in the constellation Virgo.

The most remarkable thing about the Virgo planet is that it maintains roughly the same average distance from its star as Earth does from the Sun. Therefore, its surface could conceivably be warm enough to support oceans of liquid water. These bodies of water could possibly harbor the simple amino acids that form the building blocks of life.

One should not be overly optimistic about prospects for advanced life on the planet near 70 Virginis. Its mass is thought to be enormous—between 7 to 9 times that of Jupiter. Because of its great bulk, its immense gravitation would likely prevent life from ever having the opportunity to form. Some scientists even doubt that the object near 70 Virginis is a planet, and suggest instead that it is a type of failed star called a brown dwarf.

Since the discovery of 47 Ursae Majoris and 70 Virginis, Marcy and Butler, along with several of their colleagues, have announced their finding of a half-dozen additional worlds of Jovian (Jupiter-like) proportions, circling the same number of meticulously observed stars. A veritable planet boom has begun; in 1996, new sightings were announced almost every other month. Unfortunately, because of the limitations of present-day equipment, Earth-size planets likely to harbor life have yet to be found.

To find life in space we must continue our quest. Now that we have found several extrasolar planets, we need to map out the potentially habitable worlds in our section of the galaxy. We should establish which of these bodies could support advanced life, perhaps even intelligent existence. The prospect of finding such worlds is what makes projects such as the Astronomical Studies of Extrasolar Planetary Systems so exciting.

Planet hunting is a challenge. Many planetary contenders have come and gone, dismissed for lack of sufficient proof. This difficulty in finding viable candidates probably has little to do with rarity. In fact, current theories of planetary formation suggest that planets are quite common.

Indeed, years ago it was thought that the creation of planets was an uncommon event, caused by a cosmic cataclysm such as the near collision of two stars. It was believed that during such an occurrence, material would be drawn from one of the suns. This matter would eventually solidify into planets. Because such catastrophes would happen so infrequently, there would be few planets in the galaxy. We would therefore have to attribute the existence of Earth, and of all life as we know it, to pure luck. The hope of discovering worlds orbiting nearby stars would be minimal.

Scientists now believe that planets form as natural byproducts of the process of star creation, rather than as catastrophic occurrences. According to contemporary theories, stars, along with their own planetary systems, begin their lives as protoplanetary disks: rapidly spinning clouds composed of diffuse gases, scattered dust, and large chunks of frozen chemicals, such as ice. While the dense centers of these disks eventually coagulate into stars, the material fragments lying in the disks' peripheries are gradually brought together by gravity—ultimately forming planets.

This modern picture of planetary formation implies that planets should be commonplace in the cosmos. In most cases, whenever a star is created, planets ought to be fashioned from the remaining debris. Because there are billions of stars in the Milky Way, there should be billions of worlds as well.

Then why is it fairly easy to find stars and extremely difficult to spot planets? The reason has mainly to do with their relative intensities. Under most circumstances the light of a star is much more brilliant than the radiation of a planet that orbits it. For visible light this ratio is often more than a hundred million to one. Thus, the chances of observing a planet directly with an optical telescope (an ordinary telescope using visible light) are comparable to the odds of locating a firefly in a million-acre forest fire.

By analogy, we might turn the tables and imagine a situation in which aliens situated at various distances from Earth attempt to establish that our planet exists and, further, that it is habitable. Suppose a space creature, living on the Moon, were to gaze at the Earth with a telescope. It would be obvious to the "Moonling" from its observations that the Earth is an inhabited world. (The Moonling, accustomed to seeing the Earth in the sky, wouldn't have to use a telescope to prove merely that it exists.) Signs of human creation would abound including indications in architecture, such as the Great Wall of China, as well as in agriculture, such as patchwork fields of grain that change color with the seasons.

Likewise, a space being living on Jupiter would have no trouble establishing that there exists a planet Earth. In the Jovian skies (if they were clear, that is), the Earth would stand out like a blue crystal in a black velvet case. Because Earth is blue, the being might well suspect that it has an oxygen-rich atmosphere. For us this is a sign of habitability; what such a being would make of this fact would depend on its own constitution. Because of the distance, however, it would be virtually impossible for the creature on Jupiter to discern any direct evidence of life on Earth, let alone indications of intelligence.

From the vantage point of Pluto and the outermost reaches of the Solar System, Earth would be detectable only with a strong telescope. It would take a concerted effort to look for it and, even then, it might be missed for a long time due to its diminutive size. Certainly, at that distance, no signs of life on Earth could be seen directly.

Finally, consider the lot of an extraterrestrial, living near the giant star Sirius, trying to detect Earth by using an optical telescope. Sirius lies about 50 trillion miles from the Sun—merely a stone's throw away by astronomical standards. Nevertheless, in spite of the fact that Sirius is one of our stellar "neighbors," a Sirian would find it almost impossible to make out Earth's image. From that perspective, the angular distance in the sky between the Earth and Sun would appear to be little more than one ten-thousandth of a degree apart. This would be an extremely small interval to resolve; the Earth and Sun would seem to occupy practically the same celestial location. Moreover, because the visible light put out by the Sun would be hundreds of millions of times that of the Earth, it would be extraordinarily difficult to distinguish the latter amidst the relatively bright glow of the former. Thus the Sirian would likely never know that Earth exists. Conversely, using purely optical techniques, we would likely never know if there is a planet near Sirius.

Because of the problems with employing optical astronomy to search for extrasolar planets, scientists have turned to other methods of discovery. Some researchers have suggested hunting for planets using the infrared part of the spectrum, instead of the visible range. Infrared light is radiation that has a longer wavelength and lower frequency than visible light. Normally, we cannot see infrared rays. However, with special telescopic detectors, infrared radiation can readily be recorded and analyzed.

The advantage of infrared telescopy for planet hunting is that planets, in giving off heat into space, emit most of their radiation in the infrared range. Space is extremely cold; its background temperature is just a few degrees above absolute zero. Compared to the frigidity of space, planets are pockets of warmth; their temperatures average in the hundreds of degrees. Because their temperatures are so much higher than that of the general background, they continuously radiate large quantities of heat into space. This is similar to what happens to a hot potato when it is thrown into a pile of snow; the hot potato radiates heat into the snow. In the case of planets, this heat is conveyed throughout space in the form of infrared rays. For this reason, many astronomers feel that the future of planet hunting lies with infrared techniques.

In spite of natural advantages over optical methods, infrared planet hunting is still in its infancy. Many formidable problems need to be worked out. One of the main difficulties is that even at infrared wavelengths, stars greatly outshine their planetary companions. Special optics, designed to block out the light from central stars, have been developed to help reduce this hindrance. Astronomers hope to use them to distinguish more effectively the minute image of a planet from the hazy background of its mother star's infrared light.

Another issue concerns the "twinkle effect," in which stellar light is smeared by heat-induced ripples. Earth's atmosphere contains numerous heat zones, each of which distorts light that passes through it. This is the same phenomenon that causes stars to twinkle as we watch them in the sky. Because researchers are dealing with precise measurements, they certainly do not welcome this smearing effect. Therefore, if possible, they prefer to use space-based instruments, placed high above the atmosphere's distorting effects.

Circling hundreds of miles above Earth, the Hubble Space Telescope gleams as a mechanical jewel in the sky. With its shiny panels, intricate circuitry, and pointed transmitters, it orbits our planet as a purposeful moon—a humanmade satellite in space. Fashioned by terrestrial scientists, it bears the mark of human design at its best. With this unnatural, but familiar, celestial body, scientists aspire to find other celestial bodies that are very real, but very alien. In this manner, they hope that the Space Telescope will provide a critical link between the mundane and extraterrestrial realms.

The Hubble was launched in 1990 with the ultimate in planet hunting equipment: a faint-object camera with a coronographic finger. The faint-object camera uses electronic digital sensors to scoop up as many light particles as possible from a given region in the sky. The images generated by this camera are among the crispest produced in the history of astronomical photography. Situated in front of the camera—a boon for planet searching—is the coronographic finger, a device that blocks out the center of the camera's field of view. Astronomers find this instrument to be enormously useful when attempting to record the faint light of a planet orbiting a central star, while at the same time trying to ignore the star's bright rays.

Originally, many astronomers were hopeful that the Hubble Telescope would detect planets around a significant percentage of our stellar neighbors. Unlike the case of terrestrial observatories, stars observed by the Hubble do not seem to twinkle. There is no intervening atmosphere to cause such distortion. Therefore, any planetary systems discovered could be viewed without the blurriness characteristic of ground-based observation.

Unfortunately, observation time on the Hubble is hard to obtain. Because the Hubble is so powerful, astronomers line up for years just to get a brief chance to use it for their pet projects. It would be fantastic if there were hundreds of space telescopes, each designed for a different purpose. One could be pointed at Jupiter, a second could be aimed at a distant galaxy, and yet another directed toward prospective new worlds. But, alas, there is only one space telescope, and it is greatly overbooked.

Also, even the mighty Hubble does not produce sharp enough images for astronomers to be able to detect Earth-size worlds around distant stars. A much bigger telescope, with a smoother mirror, would be needed to perform such a feat. Because such a large instrument would be extremely expensive, there is little chance that it will be constructed in the near future.

Since it is so hard to reserve time on the Hubble, and because of even that telescope's limitations, the hunt for extrasolar planets has been mainly conducted using ground-based instruments. Astronomers have developed clever means of trying to observe planets indirectly. NASA's recently launched program, Astronomical Studies of Extrasolar Planetary Systems, funds a grab bag of these methods.

Currently, there are four major ways in which researchers are hunting for planets. Technically, these are called astrometry, spectroscopy, interferometry, and photometry. For descriptive purposes, however, I refer to them respectively as the wobble method, the speed-trap technique, the zebra-stripe procedure, and the shadow approach.

The wobble method (astrometry) relies on the fact that when a distant star has a planet in tow, the star tends to wiggle back and forth. Because of their mutual gravitational attraction, the star pulls on the planet and the planet tugs back on the star. They execute this special ballet again and again as the two orbit a common point. Compared to a planet's sweeping movements, the star's motion would be subtle, but perceptible. With fine-tuned measuring instruments, researchers hope to record the minute changes in a star's position due to the effects of an unseen world.

The speed-trap technique (spectroscopy) depends on the telescopic capturing of information about the velocities of stars. If a stellar body were encircled by a planetary system, its motion relative to Earth would be slightly affected. As the planets revolved around the star, the star's speed would increase and decrease in a predictable manner. Thus, by measuring and plotting out stellar velocities, astronomers are able to probe for signs of planetary influence.

This method of searching for extrasolar worlds is analogous to a police officer's monitoring of vehicles on a highway for speeding. Imagine if such an officer were recording the velocity of a truck rolling down the highway. Suppose the truck had a rickety old trailer loosely attached to it. Let's imagine this trailer is so unstable that it swings back and forth periodically. If the officer's velocity detector were sensitive enough, it would register changes in the truck's speed due to the trailer's swaying. If the trailer were cocked to one side, it would slow down the truck at least a little bit. This would come across the monitor as a decrease in the truck's speed. Therefore, a sufficiently clever police officer would be able to look at a velocity monitor and predict whether or not a given truck has an attached trailer. Similarly, a suitably equipped astronomical team would be able to monitor the velocity profile of a star and calculate its likelihood of being gravitationally connected to a planetary system.

The third major planet-hunting technique, the zebra-stripe procedure (interferometry), involves collecting light in space from candidate stars and their potential planetary systems. Through a series of mirrors, this radiation is allowed to interfere (merge together) with itself, producing a characteristic series of bright and dark bands. The dark "zebra stripes" are analyzed for signs of bright spots that may indicate the presence of planets. The planetary radiation, once isolated, can then be further studied for signs of life-supporting substances, such as oxygen. A European investigation, called the Darwin project, and an American study, called the Exploration of Neighboring Planetary Systems, have been proposed (but not yet funded) to carry out interferometric missions sometime in the beginning of the next century.

Probably the simplest method conceptually for finding extrasolar worlds is the shadow approach (photometry). This technique involves gauging the effects that a planet passing in front of a star would have on its brightness. When a planet orbits a sun, it briefly blocks about one-hundredth of one percent of the star's light. Astronomers hope to detect this minute effect with photosensitive instruments, such as charge-coupled diodes (CCDs) that convert light into electricity.

William Borucki of the NASA Ames Research Center has proposed looking at several thousand stars per year for periodic dips in light intensity, hoping to discover dozens of Earth-size planets. NASA is currently deciding whether to fund a hundred million dollar program, called the Kepler Mission, headed by Borucki and based on his proposal. (His proposal was originally called FRESIP, for Frequency of Earth-Size Planets.) If the Kepler Mission is approved, it will begin early in the next century.

Since the 1970s, a number of researchers have been applying the first two methods—the wobble and speed-trap approaches—in their quest for extrasolar worlds. Just as one could guess whether or not a dog has fleas by the way it squirms and the way it walks, these pioneering scientists are trying to tell if certain stars have planets by the way they wobble and the way they move across space.

Astronomers expect to employ these powerful techniques in mapping out hundreds of new worlds. Already, astronomers Michel Mayor, Didier Queloz, Geoff Marcy, and Paul Butler, along with their colleagues, have used the speed-trap method to discover a number of extrasolar planets. With a combination of approaches, other intrepid planet-hunters, such as George Gatewood of the University of Pittsburgh, Bruce Campbell of the University of Victoria, and David Latham of the Harvard–Smithsonian Astrophysical Observatory, hope to reveal a host of additional bodies around neighboring suns. (Gatewood has already announced success.)

Now that the existence of extrasolar planets is fully established, scientists are trying to determine which planet supports life. This habitability test is the critical second phase in NASA's planet exploration mission. To prepare for this stage, theorists have proposed spectroscopic (analyzing light spectra) ways of testing for atmospheric oxygen content. The motivation for this analysis derives from the fact that most of Earth's oxygen is produced by plant life. Therefore, if an extrasolar world were found to have a substantial quantity of oxygen in its atmosphere, its chances of harboring life would be excellent.

Scientists need to be cautious in their underlying assumptions, however, when they decide whether or not a planet has life. Ultimately, they must come to terms with the fundamental question: what does it mean for something to be alive? Do living things necessarily require carbon and oxygen? Might life forms be based instead on other elements?

Stanley Weinbaum, in his classic story "The Martian Odyssey," pictures pyramid-shaped creatures surviving on a diet of silicon. In Fred Hoyle's tale "The Black Cloud," intelligent beings composed of pure energy seek contact with Earth. Might it be the case that extraterrestrials, instead of being creatures of flesh and blood, are life forms of radically different composition?

Indeed, a number of prominent scientists now believe that "alien microbial hordes"—colonies of microorganisms adapted to exist without light—might exist within the interiors of planetary bodies.[4] Driven by new evidence of dark ecosystems within Earth's interior, they speculate that other worlds (Europa, a large moon of Jupiter, for example) might similarly contain hidden life forms. Rather than requiring starlight—they wonder— might a considerable portion of life in the universe be buried?

Although these questions now seem academic, they may become of vital interest soon enough. Earth is becoming more and more crowded. Within a century or so, at present rates, it will reach the limits of its sustainable growth. At that point humankind will pine for the days when Earth was comparatively pristine with enough room for all. But it will, perhaps, be too late by then to salvage our spoiled planet.

Faced with rampant overpopulation, our descendants may wish to abandon their mother world and try their luck in space. Indeed the planets found by astronomers during the next few decades may comprise the future dwelling places of our great-grandchildren. This is why programs such as NASA's studies of extrasolar planetary systems are so critical. The information gleaned by such research will enable us to find alternatives to Earth if someday necessary.

Imagine our future in an age of interstellar exploration. Someday our race will no longer cling to Earth's mantle but will roam free throughout the cosmos. Vast space arks, bearing the most adventurous of souls, will be launched into the beyond. These ships will be specially designed to maintain life for hundreds of generations— enough time to reach an alien world suitable for colonization. Plants on board will freshen and replenish the air, and, along with select animal species, provide sustenance for thousands of passengers.

In all probability, the circuitry, guidance systems, and

life-support devices of the ark will be hooked up to a formidable computer. Its electronically stored atlases will carry the names and locations of potentially habitable planets—alien earths discovered perhaps by Mayor, Queloz, Marcy, Butler, and their successors. As the spaceship approaches one of these worlds, the computer will assess its potential as a new homeland for humankind. Only if conditions seem amenable—or no superior planets can be found—would a landing be attempted.

Perhaps the worlds claimed by future explorers will represent new Edens, oases in space filled with incredibly luxuriant flora and fauna. New civilizations may take root in these abundant habitats, extending and enriching the saga of the human race. Our time on Earth may be viewed someday as a drab prelude to a great age of untold material and spiritual wealth.

Or maybe the planets found will be bleak worlds, bereft of life due to unfortunate circumstances. Perhaps there will be no choice but to attempt settlement on one of these desolate globes in space. But all would not be lost in that circumstance. These worlds, though unable to support the presence of living creatures, may be of value for their mineral content. Possibly terraforming, the process of transforming a lifeless domain into a habitable one, will be mastered by then and used to open up barren worlds for colonization. Researchers have developed detailed terraforming scenarios, describing how barren planets, such as Mars, might be turned into oxygen-rich, human-friendly environments. These transformational methods may someday be applied to suitable extrasolar worlds causing, in essence, flowers to bloom in the desert.

Finally, it may be the case that we will be surprised by the presence of extraterrestrials dwelling on the worlds to which destiny takes us. Although first contact will likely be shocking to the human species, gradually we will become used to the idea that we are no longer alone in the cosmos. Though we hope these interactions between human and alien will be friendly, we must be prepared for the possibility of confronting hostile forces who refuse all attempts at communication. Sadly, it might even be the case that we would not be able to comprehend these newly discovered beings at all, or they us.

This may all seem extraordinarily hypothetical. And it is, for now. Interstellar travel is likely many decades away, possibly even centuries in our future. Unless we broaden our quest for habitable planets, we may never voyage from terrestrial shores to otherworldly realms. A good sailor needs a trusty map; a good astronaut needs a worthy guide to planetary locations.

As Frank Drake emphasizes, we can theorize away about whether or not there are inhabited worlds in space. We can debate to our hearts content about the possible nature of extraterrestrials. But until we take measures to determine what is actually out there, we really don't know a thing.

Fortunately, the first steps toward these goals have indeed been taken. Thanks to Alexander Wolszczan, we now know that planetary bodies are not unique to our own Solar System. Thanks to Michel Mayor, Didier Queloz, Geoff Marcy, Paul Butler, and others, we realize planets circle stars similar to our own Sun. Courtesy of NASA, a number of well-funded research groups are now probing for extrasolar worlds that resemble Earth. New equipment, especially the Hubble Space Telescope, has greatly increased their chances of success. The interplanetary adventure is just beginning, and where it leads will certainly be exciting.

Chapter 1
RED NEIGHBOR

> *No one would have believed in the last years of*
> *the nineteenth century that this race was being*
> *watched keenly and closely by intelligences*
> *greater than man's and yet as mortal as his own;*
> *that as men busied themselves about their various*
> *concerns they were scrutinised and studied,*
> *perhaps almost as narrowly as a man with a*
> *microscope might scrutinise the transient*
> *creatures that swarm and multiply in a drop of*
> *water ... No one gave a thought to the older*
> *worlds of space as sources of human danger, or*
> *thought of them only to dismiss the idea of life*
> *upon them as impossible or improbable.*
>
> H. G. WELLS, *War of the Worlds*

Invaders from Mars?

On a chilly night before Halloween 1938, thousands of panicked individuals packed up their belongings and fled from what they believed was an ongoing invasion from Mars. Fearing flying saucers and death rays, they raced their cars down darkened American highways, hoping to find a safe haven, free from Martian control. All the while, those still at home remained transfixed by their clunky radios, listening in horror to minute-by-minute updates of alien advances and conquests. They prayed fervently that American troops could muster enough firepower to defeat the hideous encroachers from outer space before these creatures could conquer Earth.

The news broadcasts were fictional, of course—the twisted inspiration of young writer/producer/broadcaster Orson Welles. He based them on H. G. Wells' famous novella *War of the Worlds*, changing the locale from England to New Jersey, and the format from narrative to news broadcast style. (The foreboding opening lines of *War of the Worlds* are quoted at the beginning of this chapter.)

Although Welles did his best to make his broadcasts sound as realistic as possible, he could hardly have anticipated the mass panic that ensued. Cultural historians today think that it was pre–World War II jitters that helped whip up the hysteria. The public was primed for an invasion from Europe—or somewhere—and momentarily believed Martians to be part of the same general threat.

Why Mars, though? Would the broadcasts have been as powerful had they referred to hordes of cone-shaped Saturnians or armies of bestial Jovians? Almost certainly not. Belief in intelligent life on Mars was commonplace during the 1930s; the expectation of extraterrestrial life in other parts of the Solar System was comparatively rare. Mars was widely held to be special among planets. It was thought to be a much less comfortable, but nevertheless life-sustainable, version of our own world.

Historically, these notions are fairly recent. At the time of the Mars scare, it had been only a few centuries since belief in intelligent beings on planets other than Earth had become acceptable. In Europe, before the late

Renaissance, it was considered heresy even to suggest humans were not the lone (corporeal) bearers of conscious thought. As late as 1600, Italian monk Giordano Bruno, a prominent advocate of life on other planets, was burned at the stake for his "blasphemous" teachings. Therefore, from the Middle Ages to the early Renaissance, mainly because of the danger of official condemnation, there were few vocal advocates for the existence of life in space.

After the 1960s and 1970s, scientists came to realize, ironically, that the medievals were at least partly correct. Space probes sent to other planets in the Solar System provided no indication they might harbor complex life, let alone intelligence. Humans were indeed the sole intelligent occupants of the entire Solar System, not just Earth. If advanced extraterrestrials existed, scientists concluded, they must live on faraway extrasolar worlds, not on relatively proximate Sun-orbiting bodies.

Wedged in between those bookend periods was an interval of some three centuries when it was possible, and often fashionable, to be an educated believer in beings dwelling on our immediate planetary neighbors. This interest in extraterrestrials began in the early 17th century when Italian astronomer Galileo Galilei proved with his simple telescope that the planets are not mere points of light in the sky. Rather, he showed they were solid bodies—worlds in their own right—that might potentially be walked upon by living beings. Later that century, French astronomer Bernard de Fontanelle made the bold statement:

"The Earth swarms with inhabitants. Why then should nature, which is fruitful to an excess here, be so barren in the rest of the planets?"[1]

Other noted scientists, in the centuries to follow, engaged in similar speculation. In the late 18th century, renowned English astronomer William Herschel declared that all of the planets could possibly harbor life. He even thought the Sun to be "richly stored with inhabitants."[2]

The belief in life on *all* of the other planets was rather short lived, however. As telescopic methods improved, astronomers came to realize that if there was life on other known worlds it would most likely be on Mars. They established that the other planets in the Solar System were probably either too hot or too cold to support life.

Mars, though it is farther away from the Sun that we are, was long believed (erroneously) to have only a slightly cooler climate than ours. It has four seasons, white polar ice caps that melt in summer and a thin atmosphere (now known to be exceedingly thin). Regions of its surface are covered with darkened blotches, once thought either to be seas or patches of vegetation. Therefore the red planet was at one time considered to be an especially good candidate for extraterrestrial habitation. William Herschel himself thought life on Mars to be quite similar to that on Earth. Of all the planets, he believed existence on Mars would seem the most familiar.

Interest in Mars as a potentially inhabited world grew considerably during the late 19th century with the discovery of dark bands criss-crossing its surface. These markings turned out to be an illusion, but for several decades a segment of the astronomical community thought that they were water-bearing canals, possibly used for irrigation.

The Canal Controversy

Perhaps no other debate in observational planetary science captured the public's interest as much as the Martian canal controversy. Resulting from an unlucky series of astronomical blunders and distortions, this dispute began in the late 19th century and lasted well into the

20th. For years, it placed Mars at the forefront of the popular imagination as a planet supposedly harboring a mysterious civilization.

In 1877, a time when Mars was in favorable opposition (close to Earth in its orbit and opposite to the Sun in the sky; an ideal time for observation that occurs once every 15 years), the esteemed Italian astronomer Giovanni Schiaparelli scrutinized the red planet with his 8.6-inch telescope. With his keen eye, he noted the presence of a complex network of fine dark lines, which he referred to as *canali*, or channels. He assumed that these represented natural phenomena, but the word he used was popularly translated in the English-language press as the term for artificial waterways: "canals."

For historical reasons, the idea of artificial waterways on Mars was readily accepted. The late 19th century was a magnificent time of canal building around the world. Great sea-linking canals were built across the isthmuses of Suez and Panama. Naturally, when the public first found out about the discovery of "canals" on Mars, many thought that they were the very emblem of an advanced industrial civilization. They pictured a race of machine-wielding diggers, excavating the Martian soil and building a vast network of waterways.

Schiaparelli neither promoted nor denounced this image of Martians as canal builders. His voiced opinions represented expressions of curiosity rather than of conviction. At first he thought that the *canali* were likely the product of geological forces—fault-lines, perhaps. Later he was struck by their sharp linearity and baffling complexity. In the end, he deferred to other scientists to render a verdict on this matter. In short, he did little either to quash or fan the flames of the Martian canal controversy. It probably wouldn't have become an issue of such consequence if it weren't for the persistent rallying calls of amateur stargazer Percival Lowell.

Percival Lowell was a man of fine Massachusetts stock—a genuine Boston Brahmin. Born there in 1855, his patrician ancestors were of such prominence they had cities of their own named after them: Lowell and Lawrence. As a young, highly intelligent blue-blood, he naturally found himself a place at Harvard, and, in general, followed the path that men of his social class were supposed to take. He played polo, traveled around the world, and hob-nobbed with the fabulously rich and famous. If it weren't for his abundant interest in scientific discovery and his hard-working ethics, he could easily have been just another entry in the society column. But he chose instead to pursue a scientific career and a most intriguing one at that.

Lowell's overriding interest was in the planet Mars, particularly in theories of Martian habitation. He was fascinated by the work of Schiaparelli and believed strongly that the canal issue required a satisfactory explanation. In 1892, when he found out that the Italian astronomer was no longer working on that question due to poor eyesight, he decided to take matters into his own hands. He dug into his deep pockets and pulled out enough cash to fund an observatory dedicated to the study of Mars.

Lowell decided to locate his astronomical observatory (now known as Lowell Observatory) in Flagstaff, Arizona. This was a rather bold and surprising move, but one that made a lot of sense. Arizona, at that time, was still a territory, part of what we would call the Wild West. It was more of a haven for rugged frontiersmen than for genteel scientists. Because of its dry desert climate and low population density (little industry, few gas lamps, etc.), its skies were extraordinarily clear. This clarity provided a perfect situation for stargazing. (Today, encroaching light pollution from cities such as Phoenix and Tucson has somewhat altered this ideal picture.) Lowell realized that the advantages of good telescopy greatly outweighed the disadvantages of scientific isolation. For his novel decision to situate his observatory in such a barren region, he has been commended throughout the years.

Like 1877, 1894 was a year in which Mars was in favorable opposition. One of Lowell's major goals was to finish completing his observatory by then in time for detailed studies of Mars. To his delight, his project was finished on time, and he could take up station at his newly mounted 18-inch telescope.

In June 1894 Lowell began the most thorough study of Mars to that date. With his telescope, he and his experienced assistants William Pickering and Andrew Douglass systematically combed the surface of the red planet, night after night, noting every change in its appearance that might point to the existence of sentient beings. Over time, they made thousands of sketches, detailing numerous alterations in the planet's image.

Soon after its foundation, the team began to report intriguing new results about Mars. First of all, they described a vast network of canals, far more extensive than even Schiaparelli had depicted. By the time the 1894 opposition was over they had seen almost 200 canals, over twice as many as Schiaparelli had related. (In Lowell's entire career, he charted some 700 canals.) They also noted the appearance of 53 nearly circular dark spots, which they called "lakes," and 30 dark triangles, which they referred to as "carats." They found "rivers" and "seas" and confirmed seasonal color changes that had been discovered earlier. Finally they reported the presence of blue bands around the Martian polar ice caps.

In 1895, Lowell felt that he and his assistants had gathered enough evidence of Martian intelligent habitation to present in a popular exposition. Simply entitled *Mars*, it was the first of three books that Lowell was to write on the subject. *Mars* put forward a daring theory of alien life on Earth's neighbor. Lowell postulated there had been an ancient, highly advanced civilization on Mars that eked out a meager existence amidst an extremely hostile environment. This society built the canals as a complex irrigation system, a way of conserving sparse planetary water found near melting polar ice caps.

Further evidence of Martian agriculture, Lowell maintained, manifests itself in the periodic color changes seen in the planet's visage. Just as terrestrial foliage turns from green to brown and then green again, Martian plant life, he believed, displays similar seasonal variations in hue.

Although Lowell never intended to say that Martian life in any way resembles human life, the popular press saw fit to jumble his words. Periodically they reported either that Lowell had proven the existence of men on Mars, or conversely admonished that Lowell was a lunatic for believing in such. In this fashion, Lowell's theories, twisted beyond recognition, supplied ample fodder for yellow journalism.

The Martian controversy inspired fiction writers as well. Soon after the publication of *Mars*, a host of speculative works appeared, purporting to describe life on the red planet. *War of the Worlds*, published in 1898, is certainly the best known of these. Close behind in familiarity is Edgar Rice Burroughs' series of Martian tales, which began in 1912 with the publication of *A Princess of Mars*.

Burroughs' stories tell us much about the public's view of Mars at the turn of the 20th century. He describes green-skinned warriors and copper-toned royalty who stand watch over a slowly dying world. In this fictitious Mars, resources are so scarce that they are viciously fought over. Breathable air, a precious commodity, is brought into the cities through gargantuan pumping stations. Drinking water, arriving via canals, is similarly rationed. All in all, life there is so demanding that only the hardiest of souls can survive its miseries. For this reason, if a child is born deformed he is immediately shot to death to prevent him from becoming a burden.

"I do not mean that the adult Martians are unnecessarily or intentionally cruel to their young," Burroughs wrote. "But theirs is a hard and pitiless struggle for existence on a dying planet, the natural resources of which have dwindled to a point where the support of each additional life means an added tax upon the community into which it is thrown."[3]

Much of Burroughs' prose bears the Lowellian imprint. It largely reflects, albeit in an exaggerated fashion, the view of Mars that Lowell had advanced in his theories, and that had so captivated general audiences. Much of the public had become convinced by then that an ancient race of aquatic engineers was either surviving tenuously on the planet's red soil, or else had become extinct sometime in the Martian past. Burroughs took these widely held beliefs and turned them into a comprehensive mythology.

Ironically, by the time *A Princess of Mars* appeared, scientific interest in Mars as a planet inhabited by intelligent beings was starting to decline. Aside from Schiaparelli, Lowell, Lowell's assistant Douglass, the iconoclastic French astronomer Camille Flammarion, and a few others, no one could detect the Martian canals. No matter how much they strained their eyes and adjusted their telescopes, the vast majority of astronomers could see no indication of such a network. The astronomical community began to doubt seriously whether these linear markings existed at all.

Furthermore, new estimates of the red planet's average temperature and overall atmospheric composition painted a rather grim picture of its environmental conditions. In 1907, respected British scientist Alfred Russel Wallace, Darwin's collaborator on the theory of evolution by natural selection, published a scathing review of one of Lowell's books. Wallace's critique demonstrated how Lowell's notion of a near-temperate Mars simply could not be true. For one thing, he showed temperatures were below the freezing point of water almost everywhere. Also the air was much thinner than Lowell had suggested. And finally, Wallace did a calculation which proved that any Martian effort to transport water from the ice caps to the rest of the planet via canals would have ended in failure due to rapid evaporation. Intelligent Martians, he asserted, if they did exist, would not have even attempted such a folly.

By the 1910s, few respectable scientists believed in canals, or, for that matter, in advanced forms of life on Mars. Even Lowell's assistant Pickering had long before left the group and become a vocal critic of the canal hypothesis. Lowell was crushed. Until his death in 1916 he vehemently defended his thesis and tried again and again to convince the public about the wonders of alien engineering. But belief in Martian life had evaporated.

As astronomers search today for life in other planetary systems, the Martian canal controversy provides us with a valuable lesson about trying not to jump too soon to conclusions. The human mind has the uncanny ability to create patterns where none exist. The study of visual perception is replete with examples of optical illusions. We must be wary of these as we look for regular designs that might indicate extraterrestrial intelligence. Just because a planet possesses some of the conditions for life, it doesn't necessarily meet all of the requirements. In cases in which we suspect that life might exist, we must adopt a conservative "wait and see" approach until all of the facts are in. Otherwise, the science of exobiology (the study of alien life) will degenerate into pure hype. Fortunately, since the time of Lowell, most planetary scientists have followed a sober course, and made prognostications only when they have been certain.

Mariners and Vikings

In the half-century that followed Lowell's death, his scenario for life on Mars faded rapidly from the annals of scientific literature. Yet in the popular imagination it was still virtually as fresh as ever. In book after book, film after film, television show after television show, Mars was seen as a somewhat bleaker Arizona. If you'd like to travel to the red planet, these popular presentations seemed to suggest, it would do you well to pack a warm blanket, a spare water jug, and perhaps an emergency canister of oxygen, but otherwise you'd be perfectly fine.

This public perception was to remain until the 1960s and 1970s, when a series of space missions provided overwhelming evidence that although the conditions on Earth's neighbor might not completely exclude the existence of microbes, they would almost certainly prove hostile to anything more complex. The missions revealed that the fourth planet from the Sun is uninviting for life. Possessing a sparse atmosphere with little oxygen content that nevertheless produces harsh winds and colossal dust storms, it is a fierce world indeed.

The first fly-by of Mars was performed by the Mariner 4 spacecraft in July 1965. The 22 photographic images beamed back to Earth by Mariner's radio transmitter proved most disappointing for those hoping to see a living world. The Martian surface seemed lunar in character, pocketed with myriads of craters. Newspapers reported sadly that Mars appeared to be a dead planet.

The fourth planet's landscape proved to be much more interesting, however, than Mariner 4 seemed to indicate. Further unmanned fly-by expeditions, undertaken by Mariners 6, 7, and 9, from 1969 to 1971, revealed a world of visually stunning imagery and immensely rugged topography. Mars was shown to have mountains much larger than Everest and gorges far deeper than the

Grand Canyon. The surveys demonstrated that volcanic activity and tectonic (mars-quake) activity continue to play an important role in the reshaping of Martian surface features. Therefore, Mars may not harbor complex forms of life, but it is certainly not geologically extinct.

Much to the chagrin of any remaining Lowellians, the Mariner missions found absolutely no evidence of canals. Nor did it detect any linear markings or other geometric features that might have been mistaken for such artificial waterways. Apparently what had been seen by Schiaparelli, Lowell, and others as the Martian canals were either optical illusions or transient surface patterns that somehow had mysteriously vanished.

Not everything found by the Mariners painted Mars as a lifeless planet. Geological structures were discovered that were almost certainly produced through the process of water erosion and sedimentation. Features resembling dry riverbeds, gorges, shorelines, and islands were found in abundance. These findings led many scientists to believe that liquid water once flowed along the Martian surface. The Lowellian theory of canals was not vindicated, however. If there was plentiful water, rather than passing through straight canals, it likely flowed through winding rivers to spacious seas.

Assuming that Mars once contained rivers and seas, there is little trace of their content today. Martian soil is extremely dry, making the Sahara seem like the Everglades in comparison. At some point in the red planet's history, the bulk of its water supply disappeared. Although there is still a trace of water in its atmosphere (0.03%), there is virtually none on its surface. The atmospheric water vapor is barely enough to form patches of early morning fog in the Martian valleys and wisps of clouds in its skies.

The Mariner missions whetted the appetites of scientists for more information about the possibility of micro-

Figure 1. The Mariner 4 spacecraft, shown here in NASA's Spacecraft Assembly Facility, enacted the first fly-by of Mars in July 1965. The photographs that it beamed back to Earth, depicting vast deserts pocketed with large numbers of craters, proved disappointing to those hoping for life on the red planet. (Courtesy of NASA)

organisms in the Martian soil. If Mars once had surface water, they wondered, perhaps it still harbored rudimentary life forms. In order to resolve this question by testing the planet's soil, researchers realized that a spaceship landing, not just a fly-by, would be necessary.

NASA originally planned to send a manned mission

to Mars to collect soil samples. After studying the matter they soon determined that such a voyage would be too risky and too expensive. A manned trip would take numerous months to complete and cost many billions of dollars. If any vital systems failed, rescue would be virtually impossible. Therefore, NASA decided to launch a ship with a robot probe instead.

Two Viking spacecraft, patterned after the Mariners, were launched in the summer of 1975. The following summer each of them went into orbit around the red planet and released surface landers. The Viking 1 lander touched down in July 1976 in a region known as Chryse Planitia, the "Land of Gold." The next month, the Viking 2 lander reached an area called Utopia Planitia. While the landers conducted a detailed survey of the Martian terrain and sent back daily weather reports, the orbiters broadcasted to Earth over 52,000 images of the Martian surface. Not surprisingly, Mars was found to be extremely cold, with temperatures in its *tropical* regions (where Viking 1 was situated) typically ranging from a low of minus 125 degrees Fahrenheit at night to a high of minus 20 degrees Fahrenheit during the day.

The main purpose of the Viking missions was to test for organic compounds in the Martian soil. To accomplish this goal, remote-controlled robot arms attached to the landers carefully scooped up soil samples for analysis. Several experiments were then performed to ascertain these samples' chemical and biological makeup.

In one, called the gas exchange (GEX) test, a watery solution containing organic nutrients was trickled over a sample to probe for reactions. The mixture, full of nourishing ingredients to make any known terrestrial microorganism quite happy, was nicknamed "chicken soup." After the solution was introduced to the soil, an instrument checked for the presence of life-generated gases.

Figure 2. Mars, as photographed by the first Viking orbiter. (Courtesy of NASA)

A reaction did occur—but not of the type that would have been produced by microbes. Oxygen was released from the soil through a chemical, rather than biological, process. Martian soil was found to be extremely efficient in chemically converting water vapor into oxygen.

This oxidation process was ascertained to be so strong that it would suppress life rather than foster it. The oxygen present tends to interact with metals, turning

Figure 3. Seen in this photograph is a panorama of the Martian plain as taken by the Viking 2 lander. Note the hodge-podge of boulders that litter its surface. (Courtesy of NASA)

them into metallic oxides. Unfortunately, these oxides are known to provide a poor environment for life.

We can see the pervasive presence of metallic oxides in the fiery Martian visage. Mars is red because of abundant quantities of oxidized iron (rust) in its soil. Thus the "red planet" is really the "rusty planet."

In a second test, called the labeled release (LR) experiment, a set of radioactively labeled compounds were dripped onto a soil sample. These radioactive tracers were introduced in order to scan for known biological processes. It was hoped that familiar life processes could be traced through their radioactive presence.

Once more, instead of organic reactions, a powerful chemical oxidant kicked in. Rather than indicating the presence of living organisms, this effect showed once again the all too strong—and life hindering—oxidizing properties of the Martian surface chemistry.

As a further test, in each case, the collected soil was chemically analyzed and weighed using an instrument called a gas-chromograph mass spectrometer. This device meticulously sifted through the soil for evidence of organic matter native to Mars. Although this test was extremely accurate—to one part in a billion—no trace was found of natural organic substances.

Researchers now realize that the Martian environment is self-sterilizing. Ultraviolet radiation from the Sun continually pours down on the planet, killing off the potential for life to develop on its surface. Moreover, its soil is so arid that there would be no source of sustenance for living organisms. Life has less chance of thriving in the Martian soil than it would in a fully blasting, high-temperature microwave oven on Earth.

Even in the face of these dismal findings, a few intrepid scientists still assert that there is life on Mars—far below the surface. Hope for underground life there has been buoyed by the recent discovery on Earth of bacteria residing in rocky formations miles below ground level. In 1995, biologists Todd Stevens and Jim McKinley of the Pacific Northwest Laboratory in Richland, Washington, found samples of these unusual microbes in deep aquifers near the Columbia River. These bacteria have spartan diets to say the least, thriving merely on basalt rock and whatever water they can find. According to Stevens, the same basic ingredients present where these bacteria live should exist as well in subsurface Mars. Therefore, if there is indeed Martian life, it is probably as simple and nondemanding as the subterranean basalt-eating microorganisms.

Unfortunately, to ascertain whether or not there is life far beneath the surface of Mars, scientists would need to set up a drilling rig there and keep on digging until organisms were found. It is unlikely that such a rig will be hauled there any time soon. Therefore, at present, the existence of underground life on the red planet remains an open question.

Rocky Revelations

Suppose Martian probes continue to come up empty. Does that necessarily mean that there has never been life on our red neighbor? Even if Mars is dead now, could it once have been a living world, harboring, perhaps, extremely primitive organisms?

In 1996, a remarkable scientific announcement brought these questions to the fore. Reporting in the prestigious journal *Science*, a team of NASA researchers, headed by David McKay of the Johnson Space Center in Houston, claimed to have discovered relics of Martian life from long ago. In their controversial account, they detailed the finding of the first fossil record of ancient microorganisms from Earth's red neighbor.

The focus of the team's studies was a 4 pound meteorite (the size of a baseball) collected in 1984 from the
Allan Hills region of Antarctica. Catalogued as Allan
Hills 84001, or ALH84001, the rock is thought by astronomers to have been expelled from Mars some 16 million
years ago during a cometary collision. For millions of
years, the chunk meandered through space until it finally
became caught up in Earth's gravitational field. Geolo-
gists reckon that it crashed into Antarctica about 13,000
years ago.

Examining the Martian rock, scientists have found it
to have an intricate structure. It is a labyrinth of twisted
fissures, formed millions of years ago by the penetration
of water. As the water seeped in, it carried dissolved
carbonous materials, ultimately depositing them in the
form of carbonate globules (balls of carbon-containing
substances). As in the case of stalactites hanging from the
ceiling of a cave, stony formations were created by the
action of silty fluid seeping from the surface underground.

None of this is all too surprising. Water probably
once flowed on the surface on Mars. Naturally, some of it
could have trickled into the interior and formed detailed
structures, such as fissures and globules. And some of
these infiltrated segments could have easily been blasted
off into space by a comet and eventually brought down to
Earth by gravity. Thus, a Martian meteorite with ancient
signs of water penetration, and exhibiting nothing else of
interest, would have hardly been a great find in its own
right.

What is exceptional about ALH84001 is something
stranger than a mere set of cracks and deposits. Scanning
each globule of the rock with a powerful electron microscope, the NASA team noted curious patterns of tiny,
teardrop-like markings. Some round, others long and
thin, each less than 200 nanometers in length, these mark-

ings appear to have the shapes and sizes of miniature organisms. McKay's group asserts that these indeed are microfossils—imprints of long-gone life on Mars.

Chemical analysis of the alien rock lent further weight to the NASA team's bold claim. The meteorite was found to contain substances called polycyclic aromatic hydrocarbons (PAHs), oily molecules that are common byproducts of organic processes on Earth. Moreover, the carbonate globules within the chunk were seen as having black and white rims, composed in part of fine-grained magnetite and iron sulfite particles. Such iron-based compounds are often produced by terrestrial living organisms.

Without a doubt, the NASA group has presented the best set of indications to date that life once existed on Mars. Yet no one, not even the team members, would argue that the verdict is in. Alternative explanations exist for the discovered phenomena. For instance, the teardrop-shaped markings could have been produced through inorganic processes, such as unusual sorts of crystallization. Also, PAHs and iron-based compounds have been found before on other meteorites. Their existence within the rock does not necessarily point to the influence of life.

Taken as a whole, the NASA team's analysis of the Martian meteorite is certainly encouraging. Its tantalizing suggestion that life once existed on a neighboring world serves to spark our imagination in a manner unprecedented since the days of Lowell and his canals. The mere thought of life on Mars fills one with awe and delight. True, the old images of little green men with antennae have been displaced by the new possibility of little green (or whatever) microorganisms. (Conscious life may have existed there at one point, but there simply is no evidence of it.) But if there has been life—of any kind—somewhere other than Earth, then maybe we are not alone after all.

If the results of McKay and his colleagues turn out to

be true, then perhaps life is more common in the cosmos than once believed. This would auger well for the possibility of locating intelligent beings on extrasolar planets. On the other hand, if the Martian life hypothesis were proven false, then we would need to scope out the heavens for living worlds around distant suns with even more ardor. For in that case, we would surmise that our own Solar System, except for the Earth, has always been barren. Life would seem even more precious—a rare commodity possessed by only a small fraction of planets.

In the 1970s a manned mission to Mars was planned and then canceled. The latest findings about the red planet provide a strong reason to revive the idea of such a journey. To discover the truth about our neighbor, we must sift through its surface layers with great diligence, looking for signs of primitive fossils. Only when we have completed such an analysis might the Lowellian question finally be resolved.

It may well be decades before astronauts first set foot upon the rusty soil of Mars. Until then, NASA is planning a number of interim projects aimed at better determining the chances that life once existed (or still exists) there. McKay and his associates are designing a more detailed analysis of ALH84001, as well as of 11 other meteorites believed to be of Martian origin, hoping to find fossil evidence of cell walls and other microbial remnants. Meanwhile, Wesley Huntress, NASA's chief of space science, has announced updated plans to launch unmanned spacecraft to Mars every few years, bearing robot scouts programmed to scan for life signs. These craft would land in regions of Mars thought to be more favorable environments, such as dried-up river beds in warmer regions.

The first such expedition, Mars Pathfinder, launched on December 4, 1996, arrived there on July 4, 1997, and began probing for signs of the ingredients for life. The ship contained a pyramid-shaped lander, which, in turn, carried a six-wheeled rover called Sojourner. The rover

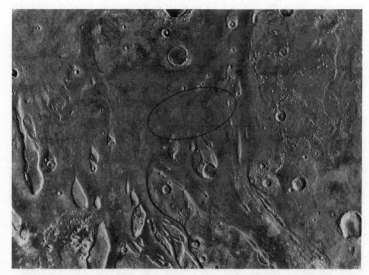

Figure 4. The Mars Pathfinder landing site, Ares Vallis, is located in the Chryse Planitia region of Mars, near the area explored by the Viking 1 lander in 1977. The landing site is indicated on the photograph by an oval. (Courtesy of NASA)

scouted out the landing site and analyzed the composition of nearby rocks. The mission found remarkable new evidence of flooding on Mars billions of years ago.

Another robot craft, called Mars Global Surveyor, was launched November 7, 1996, and is expected to arrive in the Martian vicinity in September 1997. The craft will assume a close orbit around the planet, radioing back vital information about its topography, mineral composition, and atmosphere. Over a period of months, an onboard camera will photograph the entire surface of Mars, including the sites of the Viking landings. Finally, a mission is projected for 2003 that would involve scooping up Martian soil and bringing it back to Earth for study.

These unmanned expeditions will undoubtedly augment our knowledge of Mars. Ultimately, according to Huntress, the robot missions will be limited in success, and a manned voyage to our red neighbor will likely be necessary to establish what type of life, if any, has existed there.

Earth's Other Neighbor

Mars is not the only planet in the Solar System that bears some physical resemblance to Earth. Venus, the second planet from the Sun, similarly possesses certain features in common with our world. Because both have comparable size and mass, astronomers have traditionally considered Venus to be Earth's sister planet. This image has persisted for years, mainly because, until recently, Venus's cloud cover rendered observations of its surface impossible. Like an Arabian princess of yore, the mysterious virtues of Venus were believed to be hidden behind its veils.

As in the case of Mars, speculative thinkers, as well as writers of the fantastic, long pondered the possibility of life on Venus. Its thick atmosphere was erroneously thought to be saturated with water vapor. Therefore, Venus was pictured to be a soggy world: a hot, lush, planet-wide amalgamation of England and the Pacific Northwest at their rainiest. Its drenched landscape was imagined to be blanketed with giant green foliage, perhaps akin to terrestrial species of fern.

But, alas, Venus is far from being a planetary rain forest. Space probes in the 1970s and 1980s (the Pioneer Venus and Venera series of missions) have resolutely determined that its atmosphere possesses scant water. The scorching winds sweeping over its surface are mainly composed of carbon dioxide—along with a considerable amount of sulfur dioxide and sulfuric acid.

Scientists have always believed Venus to be hot, but have only in recent years realized how hot it really is. We now know that its surface temperature would shock even Dante—from 700 degrees Fahrenheit in its highest mountain regions to a whopping 900 degrees in its valleys. Its atmospheric surface pressure would be enough to give Jacques Cousteau the bends—92 times that of Earth's at sea-level.

The fact that Venus, compared to Earth, is so close to the Sun explains only part of the reason why it is so uninviting. The sweltering climate on Venus is thought to be the product of a rampant "greenhouse effect." Its atmospheric carbon dioxide acts to reflect, again and again, trapped heat back to its surface. Thus, its surface never has a chance to cool down. Cooling could occur only through heat radiation, a process that is greatly hindered by the dense atmosphere.

Venus's condition is like that of a man sitting in a sauna, wrapped snugly in a thick wool blanket. If it weren't for its impervious covering, some of the planet's heat would have a chance to be sweated away. Instead Venus is doomed to suffocate under its wrappings.

Indeed, the fate of Venus may represent our own future. As factories and automobiles pollute our environment with waste gases, our atmosphere traps more and more heat. There is strong evidence that Earth's average temperature is increasing from decade to decade. If this tendency were to continue unabated, Earth might someday become an identical sister to Venus after all.

For now, Earth is alive, but Venus clearly is not. There is little chance that living organisms could survive the second planet's brutal climate. Space missions have confirmed that Venus seems as deserted as the Moon. If there ever was life on Venus, it has long been baked to a crisp.

Chapter 2
THE FAR REACHES OF THE SOLAR SYSTEM

The true method of investigating whether any motion can be attributed to the Earth, and if so what it may be, is to observe and consider whether bodies separated from the Earth exhibit some appearance of motion which belongs equally to all. For a motion which is perceived only, for example, in the Moon and which does not affect Venus or Jupiter or the other stars, cannot in any way be the Earth's or anything but the Moon's.

GALILEO GALILEI, *Dialogue of the Two Chief Systems of the World*

Galileo's Scope

Current knowledge of our planetary neighbors can largely be framed between two Galileos. One was the 17th century Italian astronomer, who proved that the heavenly orbs recognized by the ancients as "wandering stars" (called *planetes*, meaning wanderers, by the ancient Greeks) were genuine worlds in their own right. The other was the space probe, named for the astronomer, recently sent to Jupiter to gather vital information about the giant planet. In between have been hundreds of space missions, millions of photographs, telescopic measurements, and other analytical readings designed to probe the deepest mysteries of our neighbors.

What these probes and readings have indicated is disappointing to those hoping for life on other worlds. The Martian rock controversy aside, not a scintilla of evidence has been found so far that living organisms currently exist outside of Earth. This is truly sad. Our planet stands as a lonely oasis in the midst of an interminable desert.

Yet even the vastest deserts have their boundaries. Even if the Solar System (aside from the Earth) is completely barren, we might hope to find new living worlds in other regions of space where distant stars extend their precious warmth.

To find these other oases, we need to become masters at locating planets orbiting faraway suns, and then determining which of these might support life. We can learn much about the task that awaits us by studying how the worlds of our own Solar System were first discovered.

Five of the planets in the Solar System—Mercury, Venus, Mars, Jupiter, and Saturn—have been known since antiquity. (If Earth is included, the tally becomes six.) Unlike the fixed stars that seemed to move in unison, these planets appeared from day to day to wander along their own paths throughout the heavenly dome. Periodically, in their motions, they even seemed to reverse course. The ancients observed their distinctive behaviors and called them "wandering stars." Little else was known about these celestial objects until the time of Galileo.

Galileo Galilei, son of Pisan musician Vincenzo Galilei, was born in 1564. As a youth, he was recognized as being talented in science and mathematics. Urged by his father to embark on a practical career, he began to study medicine at the University of Pisa. He soon realized that this field was not to his liking and left university in 1585.

Galileo's true interest was in the science of mechanics. He loved inventing things, as well as figuring out how things work. While still a student, he experimented with pendula and discovered the simple law of motion that governs their rhythmic behavior. He had also developed a balance used to find the densities of materials. These and other innovations had won him considerable renown. Though he had not completed his formal education, he received offers of university positions, first at Pisa, and then at the University of Padua.

While teaching mathematics in Padua, Galileo first became aware of the invention of the telescope. In 1608, Dutch spectacle maker Hans Lippershey placed one lens in front of another and noticed that the magnification properties of the lenses were greatly enhanced. Connecting these lenses with a narrow tube, he fashioned the first telescope.

After hearing about Lippershey's invention, Galileo decided to produce his own optical instrument. He found a simple concave lens to use as an eyepiece and connected it with a lead tube to a convex lens (called the objective). With this design, he found that observed objects could be magnified by a factor of ten.

Out of curiosity Galileo aimed his newly constructed telescope at the Moon. At that time, the Moon was thought to be round and flat, like a silver coin. Much to his amazement, he found the lunar surface to be quite rugged, full of bumps, crags, dents, and fissures. The Moon, he observed, even had mountains of its own, some relatively tall. He saw it clearly to be a world of its own.

Galileo then pointed his device toward Jupiter and discovered its four most prominent moons. (Today we

Figure 5. The Moon, as seen by the Apollo 17 mission. As Galileo pointed out, it has a rugged topography, full of craters, mountains, and plateaus. (Courtesy of NASA)

know that it has 16 satellites at least.) For political reasons he named these the Medicean stars, after the powerful Medici family of Florence.

Galileo was a shrewd man who realized the power of flattery. To further bolster his influence with the Medicis he dedicated his first astronomy book, *Starry Messenger*, to Cosimo de Medici II. His strategy paid off handsomely. He was immediately appointed to the prominent position of Chief Mathematician and Philosopher in Florence.

In 1610, Galileo, in his new office began a detailed survey of the then-known Solar System. He examined Venus, noting its cycle of phases. Like the Moon, Venus periodically waxed and waned, transforming from a disk to a crescent and then back again. These variations suggested that Venus, rather than producing its own light, was illuminated by the Sun.

Galileo next turned his telescope to Saturn, noting its unusual shape. He was puzzled by its complex structure and could not come up with a satisfactory explanation. Then, examining the Sun he mapped out its patterns of sunspots. Noting that these specks seemed to circle a common axis, he concluded that the Sun rotates, whirling its spots around with it as it does.

Based upon his astronomical research Galileo drew a number of important conclusions about the planets, revolutionary for their time. First, he concluded that the planets were worlds of substance, rather than mere points of light. If Earth were seen from another part of the Solar System, it would appear to be a wandering star as well. Thus, these variations in appearance stem from differences in perspective.

Second, Galileo deduced that Earth and the other planets revolve in simple orbits around the Sun. This fact—now considered common knowledge—was heretical for its times. By strongly advocating this point of view, Galileo cast into doubt centuries of religious teachings in Europe.

The official position of the powerful Church hierarchy from the Middle Ages until the 18th century was that Earth is the center of the universe. The planets in the sky are peripheral entities engaged in a perpetual dance around our sacred domain. Earth, not those "wandering points of light," represents the focus of creation.

After observing with his telescope the basic similarities between Earth and the other planets, Galileo realized that the geocentric viewpoint advanced by the Church was untenable. Drawing heavily on the work of Polish astronomer Nicholas Copernicus, he argued instead for a heliocentric (Sun-centered) cosmology. Galileo showed schematically that the observed behavior of the "wandering stars," including their so-called retrograde (backward) motion, could be well explained by assuming that they travel around the Sun. He summarized his thoughts in an influential book entitled, *Dialogue Concerning the Two Chief World Systems: Ptolemaic and Copernican*, where he compared the geocentric scheme of Greek philosopher Claudius Ptolemy with the heliocentric notion of Copernicus. Naturally, Galileo's treatise favored the latter.

Church officials were enraged by Galileo's spurning of their established teachings about the nature of the universe. Step by step, they tried to quash his advocacy of heliocentrism. First, they made it officially illegal to teach Copernican thought. They issued warnings to Galileo not to advance such views. Finally, in 1633, they summoned him to Rome and forced him to recant. Afraid of facing the wrath of the Inquisition, he got down on his knees and begged Church authorities for forgiveness for his "errors." A broken man, he spent most of the last decade of his life under house arrest.

Requisites to Motion

Around the same time Galileo was beginning to scan the heavens with his telescope, German mathematician Johannes Kepler was discovering simple laws that describe planetary motion. Kepler, who was born in 1571, began his studies at the University of Tuebingen as a religious scholar. While at Tuebingen, he became interested in the realm of the planets. He studied the Copernican system and became convinced of its validity.

In 1600, Kepler was invited to assume a post in Prague, working with pioneering astronomer Tycho Brahe. Brahe was known for his keen visual observations of the planets in the days before the telescope came into prominence. A year after Kepler arrived, Brahe died. Assuming Brahe's position, Kepler took the opportunity to sort through the wealth of astronomical information that his predecessor left.

From this set of data, Kepler postulated three laws of planetary motion. The first is that planets follow elliptical orbits, with the Sun as one of the foci. An ellipse is a geometric figure that resembles a stretched circle. It is mathematically defined in terms of two internal points, called foci. These are designated such that if one chooses any given point on the ellipse, then that the sum of that point's distance to one focus plus its distance to the other focus is a constant.

Kepler's second law has to do with the speeds in which planets move around the Sun. Basically, it states that planets move faster along the parts of their orbits when they are closest to the Sun and slowest along the parts when they are farthest.

The third descriptive law that Kepler discovered pertains to the time that each planet takes to circle the Sun, called its orbital period. The Earth's orbital period, for instance, is one year. Jupiter, on the other hand, takes a

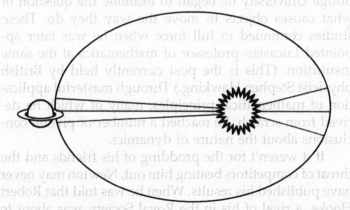

Figure 6. Kepler showed that planets travel around the Sun in elliptical orbits, with the Sun as one of the foci. He also found that planets traverse equal areas in equal times (indicated by triangular strips).

full 12 years to revolve around the Sun. Kepler found that the cube of a planet's distance from the Sun is proportional to the square of its orbital period. Thus a planet four times farther away from the Sun than the Earth would take eight times as long as Earth does to complete its orbit. (These specific numbers were chosen for their arithmetic simplicity; there are no planets four times farther away from the Sun than Earth.)

Kepler found these laws through empirical means, rather than through the application of physical principles. His conclusions were drawn from examining Brahe's data and looking for patterns. They lacked an explanation based on the laws of physics. In short, his results begged for a brilliant scientist to come along and explain *why* the planets move the way they do, not just *how* they move. The genius who performed this feat was none other than the English inventor of calculus and classical mechanics, Sir Isaac Newton.

Newton was born in 1642, the year Galileo died. Like Kepler, he studied religion, but was also attracted to science and mathematics. While an undergraduate at Cambridge University he began to examine the question of what causes objects to move the way they do. These studies continued in full force when he was later appointed Lucasian professor of mathematics at the same institution. (This is the post currently held by British physicist Stephen Hawking.) Through masterful application of mathematical principles, many of which he derived from scratch, he reached a number of pivotal conclusions about the nature of dynamics.

If it weren't for the prodding of his friends and the threat of competitors beating him out, Newton may never have published his results. When he was told that Robert Hooke, a rival of his in the Royal Society, was about to announce similar findings, Newton became enraged. He decided to publish the work considered his masterpiece,

Figure 7. Sir Isaac Newton, 1642–1727, founder of classical physics. (Courtesy of AIP Emilio Segrè Visual Archives, W. F. Meggers Collection)

Philosophiae Naturalis Principia Mathematica (Mathematical principles of natural philosophy), or *Principia*, for short. This book, which appeared in 1687, is arguably the most influential scientific text of all times.

For our purposes, we shall consider two critical sections of Newton's treatise. The first concerns his principle of gravitational attraction. Newton discovered that every massive object in the universe exerts an attractive gravitational force on every other object. For a given pair of objects, the magnitude of this force is equal to the product of their masses, divided by the square of their distance from one another, times a universal constant of proportionality. For example, the gravitational pull between two objects quadruples if either both of their masses are doubled or if the distance between them is halved.

The second part of Newton's *Principia* that is important to this discussion states his three laws of dynamics. These guiding principles explain how objects react to applied forces. Given full knowledge of an object's initial position, initial velocity, and total forces experienced, these laws dictate the manner in which it travels through space.

Newton's first law of dynamics is also known as the principle of inertia. Inertia is the propensity for an object at rest to remain at rest, and for an object moving at constant speed to stay moving in the same direction at that speed. Newton discovered that in the absence of external forces, or when external forces are exactly balanced, an object exhibits inertial behavior. In that case, if it were initially sitting still, it would never move, but if it were already moving, it would keep on going at the same speed.

Newton's second law concerns the unbalanced application of force. This principle states that if an unbalanced force is applied to a body, it consequently accelerates along the direction of the force. For example, if a baseball is struck with a bat, the ball speeds up along the path of the bat's impact.

Finally, Newton's third law is the well-known statement: "For every action there is an equal and opposite reaction." This assertion elucidates the fact that if object *A* exerts a force on object *B*, then *B* must exert a force of the same magnitude but opposite direction on *A*. Thus, the Earth's gravitational force on the Moon is equally as strong as the Moon's oppositely directed gravitational force on the Earth.

Combining Newton's theory of gravitation with his

laws of dynamics yields the extraordinary ability to predict the motions of orbiting bodies. For example, Newton's equations applied to the planets show that they
must move along elliptical trajectories, with the Sun as
one focus. The gravitational force of the Sun acts to steer
the planets along such paths; otherwise they would go
hurling out into deep space. In this fashion, Newtonian
principles provide the physical justification for Kepler's
laws.

Eyes in Space

It is a credit to the precision and power of Newtonian
mechanics that today spaceships are sent to other worlds
in the Solar System along exact trajectories dictated by
physical principles. In the case of the manned Apollo

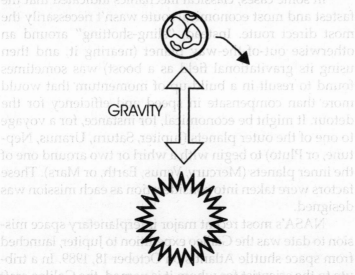

*Figure 8. As shown by Sir Isaac Newton, the force of gravity causes the Earth to orbit
around the Sun.*

missions to the Moon in the late 1960s and 1970s, astronauts trusted that Newton's laws properly applied would guide them to safety. From the instant that they were launched into space, their spacecraft followed the paths that Newtonian dynamics predicted. Otherwise, they could never have landed on the Moon and returned home successfully.

The Mariner, Viking, Pioneer, and Voyager series of space probes that surveyed the Solar System in the 1950s, 1960s, and 1970s (along with many other exploratory craft) represent additional examples of how NASA utilized classical (Newtonian) mechanics to plan out its missions accurately. Computers programmed to sift through reams of data about the probes and the planets they "flew by" or visited used Newton's laws to plot out detailed scenarios for each voyage. Based on their calculations, NASA scientists carefully chose launch times and velocities to maximize each mission's potential for success.

In some cases, classical mechanics indicated that the fastest and most economical route wasn't necessarily the most direct route. Instead, "sling-shotting" around an otherwise out-of-the-way planet (nearing it, and then using its gravitational field as a boost) was sometimes found to result in a build-up of momentum that would more than compensate in speed and efficiency for the detour. It might be economical, for instance, for a voyage to one of the outer planets (Jupiter, Saturn, Uranus, Neptune, or Pluto) to begin with a whirl or two around one of the inner planets (Mercury, Venus, Earth, or Mars). These factors were taken into consideration as each mission was designed.

NASA's most recent major interplanetary space mission to date was the Galileo expedition to Jupiter, launched from space shuttle Atlantis on October 18, 1989. In a tribute to the scientist for whom it is named, the Galileo craft delivered pivotal information about the makeup of the

Figure 9. On July 13, 1995, NASA's Galileo probe separated from the Galileo orbiter and began its entry into Jupiter's mysterious thick atmosphere. At the time of the separation, these craft were 400 million miles from Earth. (Courtesy of NASA)

giant planet, as well as spectacular photographs of the Jovian moons and atmosphere. Now considered a successful endeavor, the Galileo mission was at one time believed to be doomed to failure.

Two years after Galileo was launched, its main radio antenna refused to open properly, rendering it inoperable. Therefore, all signals sent by the craft needed to be relayed through a small auxiliary antenna. It was so difficult for NASA scientists to detect these faint messages that the data transmission rate was slowed by at least a factor of 10,000.

Nevertheless, even at this sluggish rate of exchange, the craft conveyed to Earth important new information about the composition of Jupiter's atmosphere, including revised estimates of its water content. The amount of water on Jupiter was found to be even lower than the small quantity expected. Though it may be stormy weather much of the time on Jupiter, it certainly never rains.

Initially, Galileo's results seemed to indicate that Jupiter's atmospheric helium content was less than half of the amount predicted by theory. Many scientists were baffled by this low estimate, and concluded that leading models of the Solar System's evolution must be faulty. However, re-analysis of the data revealed that readings from the probe were wrongly interpreted. Galileo's helium measuring team, headed by Ulf von Zahn of the University of Rostock in Germany, announced their reassessment in March 1996. According to their revised estimates, Jupiter's upper atmosphere is composed of 24 percent helium, close to the value indicated by theoretical models.

Galileo's rendezvous with the giant planet was the last stop on a circuitous voyage across the Solar System. After its launch from Atlantis, it began what is known as a Venus-Earth-Earth-Gravity Assist (VEEGA) phase. This is one of the methods employed by NASA to utilize the gravitational fields of inner planets to boost the speeds of outgoing spacecraft. In the first part of this route, which took place in winter 1990, Galileo "sling-shot" around Venus, then returned to the neighborhood of Earth. It swung around Earth twice, gaining considerable momentum, before heading out to the giant planet for its long-awaited encounter.

Finally, in December 1995, Galileo reached Jupiter. After assuming a steady orbit around the planet, it dropped a probe to its surface. As the probe parachuted through the thick Jovian atmosphere, it radioed valuable data out to the orbiting craft, which, in turn, relayed the information back to Earth. The probe plummeted faster and faster, until, within 75 minutes, it became a fireball twice as hot as the Sun's surface. In a short time, it transmitted its last bit of data before becoming pulverized by the immense pressures of surrounding gases.

During its brief descent, the probe witnessed a panorama radically different from vistas on Earth, Mars, or the other two inner planets. Unlike the rocky worlds of the inner Solar System (known as the "terrestrial" planets for their material resemblance to Earth), Jupiter's bulk lies in cold, swirling gases and frozen chemical ices. With its enormous internal pressures, crushing gravitational forces, frigid temperatures, and composition of mainly hydrogen and helium, with traces of methane and ammonia, it is hard to imagine a world less hospitable to life. It is safe to say that no living creature gazed up at Galileo's probe as it plunged into the heart of the Solar System's giant.

Realm of the Giants

Extending outward from the Sun, the Solar System resembles a trail of ants followed by a herd of elephants. Mercury, Venus, Earth, Mars, and their respective moons (Mercury and Venus are moonless; Earth has one; Mars has two) are small rocky orbs. Jupiter, Saturn, Uranus, and Neptune are each enormous in comparison to the terrestrial (inner) planets. One thousand Earths—or ten thousand Mercurys—could easily fit within Jupiter's spacious interior.

With Jupiter as the archetype, the large outer worlds of the Solar System are sometimes called the Jovian planets. These four giants share a similarity of composition; each is a mixture of simple gases and ices. Locked within their interiors are the primordial components of our sector of space.

The outer two gas giants, Uranus and Neptune, are relative newcomers to our picture of the Solar System. While Jupiter and Saturn are large enough and close enough to Earth to be seen readily with the naked eye, Uranus and Neptune can only be viewed easily by careful telescopic observation. (In an especially dark sky, Uranus can be made out with the naked eye, but just barely.)

Indeed until the late 18th century, popular scientific opinion held that Saturn's orbit formed the strict outer boundary of the Solar System. Much of the astronomical community believed that the number of planets circling the Sun was necessarily limited to six. The "wandering bodies" known by the ancients were erroneously thought by many to be the only ones that could conceivably exist.

Kepler, for example, in his more mystical writings on astronomy, related the number of then-known bodies in the Solar System to the number of regular (Platonic) solids. It is a proven geometric fact that there are only five of these solids, which have the special property of having the same equilateral geometric figure (triangle, square, etc.) on all sides. Finding mathematical similarities between the progression of these forms and the succession of planetary orbits, Kepler came to the conclusion that the number of planets was limited as well.

So strong was the belief that the Solar System ended with Saturn that when William Herschel discovered Uranus in 1781 with his 6½-inch telescope, he believed for quite some time that he had found a new comet. He first observed the seventh planet in March of that year while scanning an interesting region of the sky near the Crab Nebula. He observed it, day by day, as a bright spot moving slowly across the celestial dome. He reported his findings to the Bath Philosophical Society in England. When the tracked object failed to develop a tail, as a comet should when it nears the Sun, he and his scientific contemporaries began to suspect that it was a planetary body instead. Finally, after the newly found body was observed for a number of months and was shown to obey Kepler's laws, at least approximately, they realized that it must be a new planet.

There was some thought among the British astronomical community of naming the new planet "Herschel." Herschel himself wanted to name it after King George III. But good taste won out, and it was named instead after the Greco-Roman god Uranus, who was the mythical father of Saturn (who in turn was the father of Jupiter).

Uranus was not considered the outermost planet for long. Disturbances in its orbit soon led astronomers to believe otherwise. After Uranus's discovery, astronomers meticulously plotted its orbital trajectory for years to try and reconcile its path with that predicted by Newton's gravitational and dynamical principles. Surprisingly, Uranus' orbit was found to differ from prediction by a significant amount. By 1830, when this discrepancy was found to have grown to more than four times the size of the planet's diameter, researchers could only conclude that another planet, even farther away, was influencing the motion of Uranus. The search was on for a member of the Solar System more distant than Uranus.

Newtonian mechanics have their practical limitations. Though, in theory, Newton's laws exactly define the ordinary motions of objects (for high enough speeds, however, Einstein's special relativity must be applied), in practice it is difficult to deduce the behavior of more than two interacting bodies. The "three-body problem"— finding the precise paths of three bodies subject to each others' mutual forces—has challenged mathematicians for generations.

The three-body problem, applied to gravitation, can be simply stated as follows. Consider three mutually interacting objects—the Sun, for example, and two planets

Figure 10. William Herschel, 1738–1822, discoverer of Uranus. (Courtesy of AIP Emilio Segrè Visual Archives, E. Scott Barr Collection)

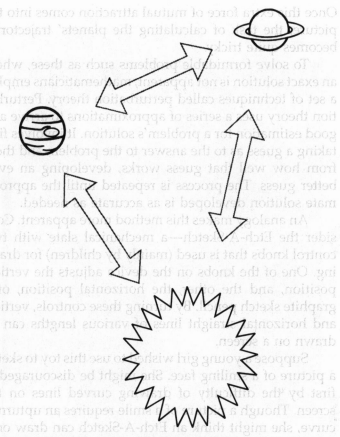

Figure 11. The three body problem—determining the motions of three objects (such as Jupiter, Saturn, and the Sun) under the influence of their mutual gravitational fields—can prove quite tricky, and is usually solved by computer.

in orbit around it. Clearly, the Sun exerts a gravitational pull on each of these orbiting bodies. Given each of their masses and radial distances, these forces can readily be computed. Their orbits might easily be determined as well, if it weren't for the additional fact that the two planets exert a gravitational attraction on one another.

Once this extra force of mutual attraction comes into the picture, the task of calculating the planets' trajectories becomes quite tricky.

To solve formidable problems such as these, where an exact solution is not apparent, mathematicians employ a set of techniques called perturbation theory. Perturbation theory uses a series of approximations to arrive at a good estimation for a problem's solution. It involves first taking a guess as to the answer to the problem and then, from how well that guess works, developing an even better guess. The process is repeated until the approximate solution developed is as accurate as needed.

An analogy makes this method more apparent. Consider the Etch-A-Sketch—a mechanical slate with two control knobs that is used (mainly by children) for drawing. One of the knobs on the device adjusts the vertical position, and the other, the horizontal position, of a graphite sketch pencil. By turning these controls, vertical and horizontal straight lines of various lengths can be drawn on a screen.

Suppose a young girl wished to use this toy to sketch a picture of a smiling face. She might be discouraged at first by the difficulty of drawing curved lines on the screen. Though a picture of a smile requires an upturned curve, she might think an Etch-A-Sketch can draw only straight lines. Eventually she might learn to develop a perturbative solution to the problem. By adjusting the two control knobs she could use a series of short straight lines to approximate a curve. Each time the "pencil" strayed from the desired image, she could turn one of the controls and etch in another line to produce a closer fit. In this manner, she could render a linear sketch that for all practical purposes would look like a curved smile.

Similarly, perturbation theory in physics can be used to model closer and closer the path that an object takes under the influence of gravity. Each time an attempted solution is found to deviate from reality, the discrepancy can be used to develop a better approximation. Eventually, after enough revisions, the object's trajectory can be well modeled.

In this manner, several 19th century mathematicians attempted to find the source of deviations in the motion of Uranus. Independently in the 1840s, two succeeded in solving this problem: John Couch Adams in England and Urbain Jean Joseph Le Verrier in France. Each predicted the existence of an eighth planet revolving around the Sun at a much farther distance than Uranus. Although Adams was first to come up with a solution, his results were overlooked until Le Verrier made his own announcement of success.

Le Verrier mailed his predictions to German astronomer Johann Galle of the Berlin Observatory and persuaded him to search for an eighth planet. Le Verrier's estimates turned out to be quite good. On his first attempt, in 1846, Galle found a new planet close to the location designated by Le Verrier. Because of its sea-blue color, the planet was named Neptune, after the Greco-Roman god of the sea.

Until recent decades, because of their great distance from Earth, little was known about Uranus and Neptune. In the late 1970s and 1980s, two Voyager spacecrafts, 1 and 2, took advantage of a favorable planetary alignment to fly by the giant outer planets. While both Voyagers swept past Jupiter and Saturn, Voyager II alone went on to Uranus and Neptune, reaching the latter in 1989.

The vital information and magnificent photographs sent back by the Voyagers throughout their 12 year mission ushered in a veritable revolution in planetary science. Never before could we witness the outer planets, in all their stunning solitary beauty, from such an ideal van-

Figure 12. The Planet Saturn as photographed in 1980 by the Voyager 1 spacecraft. (Courtesy of NASA)

tage point. From the new data received, scientists have revised their picture of the Solar System in many significant ways.

Before the age of Voyager, astronomers thought that Saturn was the only world with rings. Although Saturn's are the most prominent by far, other large planets have been found to be encircled as well. We now know that each of the gas giants has its own intricate ring structure. These rings are composed of myriad icy particles—some as small as peas, others as large as boulders—trapped in tight orbits around each planet.

The Voyager missions have shown that each planet's ring structure has a unique configuration. Jupiter possesses a single thin ring, which was discovered by Voyager 1 in 1979. Ten years later, Voyager 2 recorded Neptune's narrow set of four dusty haloes. And although a Voyager craft was not the first to spot the rings of Uranus—nine were discovered in 1977 by James Elliot of the Kuiper Airborn Observatory—Voyager 2 confirmed their existence, found two more, and helped astronomers to map out their main features.

Before the Voyager missions, Uranus was thought to have only five satellites (two of which were discovered by Herschel himself), and Neptune a paltry two. The images sent back by Voyager 2 revealed that our moon count was way too law. The craft found ten new satellites for Uranus, and six more moons for Neptune.

With these additions, the current tally for the number of moons in the Solar System has grown to over 60—and still counting. (Recall that only five were known in Galileo's day, including Earth's Moon.) These range in size from about 10 miles across, in the case of Leda, to over 3000 miles across, in the case of Titan, Ganymede, and

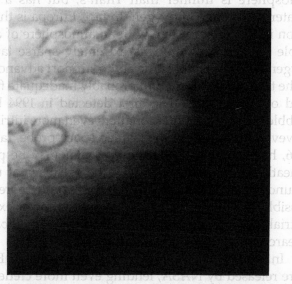

Figure 13. The Hubble space telescope's first picture of Jupiter, taken on March 11, 1991, by its wide field planetary camera. Jupiter's most prominent feature, the Great Red Spot, is just rotating out of view on the far right side of the photograph. (Courtesy of NASA)

Callisto. (Each of these satellites orbit Jupiter, except for Titan, which circles Saturn.)

Titan is the only moon in the Solar System with a thick atmosphere. Because Titan's atmosphere contains some elements that are similar to ours (an abundance of nitrogen, as well as traces of simple organic substances) some have speculated that it may also harbor chemicals that are the precursor to life. There is some chance that amino acids may have formed on Titan from its rich broth of hydrocarbons and other organic materials. The existence of such "prebiotics" on Saturn's largest satellite remains an open question. Even if these building blocks for life were found, however, no one would expect to find complex organisms on such a frigid world.

Europa, one of Jupiter's satellites, represents another conceivable locale for primitive life or its precursors. Its atmosphere is thinner than Titan's, but has a much greater percentage of oxygen. In fact, Europa is the only moon in the Solar System with an atmosphere of appreciable oxygen content. This extremely sparse layer of oxygen—certainly not sufficient to support advanced life of the terrestrial variety, and probably inadequate for any kind of surface life—was first detected in 1994 by the Hubble telescope. To make matters even more intriguing, however, images from the Galileo space probe, taken in 1996, have suggested that liquid water may be present beneath Europa's surface ice. If the existence of underground water is confirmed, Europa would represent a possible candidate for primitive subterranean extraterrestrial life. In that case, NASA would likely embark on a search for signs of organisms there.

In April 1997, startling new photographs of Europa were released by NASA, lending even more credence to the hypothesis that there is substantial liquid water beneath that satellite's frozen surface. In the most detailed pictures to date of Europa—taken by the Galileo spacecraft as it came within 363 miles of Jupiter's moon—icy domes, cracks and ridges were seen scattered throughout its exterior. The labyrinthine nature of these features indicates that an underground Europan ocean has been continuously re-shaping the moon's surface for eons and eons—repeatedly smashing ice chunks together and then pulling them apart, like cubes jostled in a blender. Recent indications that Europa has a weak magnetic field seem to confirm this picture of churning geological activity beneath its surface.

Could life exist within the murky depths of such a subterranean ocean? At least one prominent scientist, John Delaney, an oceanographer at the University of Washington, is willing to bet on the presence of aquatic life beneath the surface of Europa. But other researchers, such as Torrence Johnson, the Galileo mission's chief scientist, point out that water is only one of life's many prerequisites. He emphasizes that more information about the Europan environment is needed before conclusions can be drawn.

Europa's ocean, if it could be directly observed, certainly wouldn't look like the blue Pacific. Shielded by ice, it would more closely resemble the dark waters beneath the Arctic ice cap. Voyager images, taken in the 1980s, point to another world—Neptune—that is more visibly "oceanic" in appearance.

The crowning achievements of the Voyager missions were their distinct visual records of the outer planets—never before photographed so well. Perhaps the most spectacular of these were the sublime portraits of Neptune taken by Voyager 2. Neptune surely lives up to its name as the planet of the sea-god. Its pale blue face, marked with vast, swirling eddies, is reminiscent of the ocean on a stormy night. But it is hardly salt water that gives Neptune its characteristic color. Rather, Neptune's deep cover (composed mainly of hydrogen and helium gases) is tinctured with blue methane. Often, jets of methane are thrust upward into the cold reaches of the planet's upper atmosphere, where they freeze into puffy white clouds.

Neptune is the windiest planet in the Solar System. In its equatorial region, gaseous formations sweep west-wardly across the planet's face at speeds of up to 1,200 miles per hour. On Earth, the Sun's heat provides the main driving force behind its winds. Considering the eighth planet's considerable distance from the Sun, what, then, powers its mighty winds?

Surprisingly, the main source of energy on Neptune comes from within. Deep inside of Neptune is a turbulent core, a remnant of the planet's primordial form. This turbine supplies the power that propels the planet's highly dynamic atmosphere, constantly in a state of rapid percolation.

These powerful forces drive Neptune's Great Dark Spot, undoubtedly its most noticeable surface feature. This colossal formation—close to the size of Earth—greatly resembles Jupiter's better-known Great Red Spot. Both comprise tremendous whirlwinds in the atmo-spheres of their respective planets, formed from pressure and temperature variations. The difference in color be-tween Neptune's and Jupiter's great spots lies in the dis-tinct chemical compositions of the two gas giants.

Sometimes a blemish enhances the beauty of the wearer. In Neptune's case, the Great Dark Spot provides a striking visual counterpart to the rest of the planet's de-ceptively serene deep blue tones. It is only fitting that the Voyager missions ended with a world of such breathtak-ing contrasts.

Planet X

Pluto, the ninth member of the solar system, is the only outer planet that the Voyager craft never encountered. A fraction of the size of Earth, it is much smaller than the other worlds of the outer solar system. Pluto was first spotted in 1930 by American astronomer Clyde Tombaugh and was the last of the Solar System's planets to be discovered.

The search for Pluto was triggered by reasons similar to those that motivated the hunt for Neptune a century earlier. In both cases, orbital anomalies were found in planets already encountered. These observed disturbances were believed to be caused by the gravitational influence of unseen worlds. (In Pluto's case, as we'll see, the disturbances turned out to be experimental errors; by the time this was realized, Pluto was already found.) With this motivation, thorough searches took place for the missing planets, finally leading to success.

The quest for the ninth planet began in earnest in the beginning of the 20th century, spurred on by the enthusiastic interest of renegade astronomer Percival Lowell and his dedicated observatory staff. Lowell's belief in a world beyond Neptune stemmed from his conclusion (which he shared with many of his contemporaries) that Neptune's gravitational influence was not strong enough to entirely account for Uranus' behavior in space. He supposed that Uranus' perturbations from a simple elliptical orbit required the existence of an additional world beyond Neptune, which he called *Planet X*. Momentarily turning his attention from the question of life on Mars, Lowell directed his assistants to scan photographic plates for evidence of a ninth planet.

By 1910, Lowell's staff had examined numerous astronomical photographs and had turned up no evidence of Planet X. Meanwhile, Lowell had found out that his former assistant, William Pickering, was also in the race for a ninth planet. Pickering seemed to be devoting considerable resources to his own quest for a body he called *Planet O*.

To make matters worse, Lowell was coming under increasing attack for his Martian prognostications. The press was constantly skewering him for his beliefs in life on the red planet. Distortions of his ideas—shamelessly attributed to him—were appearing all the time. Lowell felt that he needed to do something to rescue his tarnished scientific reputation.

Lowell readied himself for battle. Hoping to snare a precious trophy for his mantle, he redoubled his efforts to be the first to find Planet X. Revising his strategy, he decided to use the mathematical theory of perturbations to predict a location for the planet. In this fashion, he hoped to repeat the strategies of Adams and Le Verrier in successfully discovering Neptune's location.

Again and again, Lowell developed detailed predictions for the position of a trans-Neptunian world. Each time, he summoned his staff and gave them the task of finding the planet. But, alas, Lowell died of a massive stroke in 1916, before he could see his efforts meet success.

Even after the death of its founder, the Lowell Observatory continued its search for Planet X. In 1929, when amateur astronomer Clyde Tombaugh wrote to the observatory staff about his studies of Mars and Jupiter, they decided to hire him to help out with their task. They saw in Tombaugh a young enthusiast with the energy and patience required to scan through countless photograph plates looking for telltale planetary behavior.

It took Tombaugh 27 hours by train to journey from his native Kansas to Flagstaff depot (the nearest station to the observatory). There, director V. M. Slipher met him and took him up to Mars Hill, where the Lowell Observatory was situated. He was introduced the next day to the new 13-inch photographic telescope, still under construction. Through his knack for astronomical instrumentation, he quickly taught himself how to use the device.

In April 1929, Tombaugh began to collect and com

Figure 14. Clyde Tombaugh, 1906–1997, discoverer of the planet Pluto. (Courtesy of New Mexico State University. Photo credit: R. Sterling Trantham)

pare images, using a photographic comparison apparatus, called a "Blink-Comparator." Unlike Galle who found Neptune in a dash, Tombaugh ran a veritable marathon before reaching success. He labored for month after grueling month on the project, looking at plate after plate, until he was wracked with utter fatigue.

Finally, in January 1930, Tombaugh informed Slipher that he had discovered the long-sought planet. Tombaugh saw it as a tiny dot, having a slightly different position on two separate photographic plates, taken several days apart. Its wandering motion marked it as a planet. After confirming Tombaugh's results, Slipher made a public announcement on March 13, the 75th anniversary of Lowell's birth, that Planet X was found.

After much deliberation, the astronomical community elected to name the planet Pluto. This name was chosen to continue in the tradition of naming the planets after Greco-Roman gods, in this case the god of the underworld. Also, the first two letters of Pluto, incorporated into the astronomical symbol for the planet, serve as a fitting memorial to Percival Lowell.

Pluto was found to have a peculiar orbital behavior. Due to its orbit's eccentricity (stretchiness), Pluto spends 20 years out of its 249 year circuit rotating closer to the Sun than Neptune. For the rest of its orbit, it constitutes the farthest world from the Sun—40 times more distant than Earth. Consequently, it has been extraordinarily difficult to analyze with a telescope. Only in recent years has the Hubble space telescope provided us detailed images of this elusive world.

Is Pluto the terminus of the Solar System or are there additional worlds beyond it? Though we now believe the former—that Pluto is the ultimate world of our system

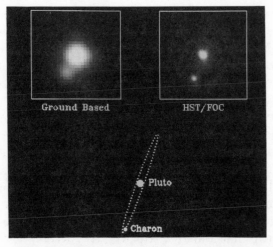

Figure 15. In 1990, the Hubble space telescope's faint object camera was used to obtain the first clear photographs of the planet Pluto and its satellite Charon. Note the sharpness of the image on the right taken by the Hubble compared to the one on the left taken by a ground-based instrument. (Courtesy of NASA)

aside from the small icy objects that form the Kuiper belt, the source of many comets—there was once substantial belief in a tenth planet farther out. This interest stemmed from an erroneous notion that there were gravitational perturbations of the known planets yet to be explained.

Soon after the announcement of Pluto's discovery, scientists realized, to their chagrin, that it was not large enough, or positioned well enough, to account for the phenomena that motivated its search. Tombaugh set out for 13 more years to find the real Planet X. Finally, after an exhaustive search of the heavens, Tombaugh concluded that there is no Planet X.

Researchers now believe that the behavior of Uranus can be fully accounted for by the gravitational effects of known planetary bodies. Neptune's and Pluto's motions are completely understood as well. The anomalous data that motivated the search for Planet X were probably just an experimental fluke. Pluto, a tiny world, has far too little gravitational influence to have caused the perceived disturbances. Nevertheless, it is fortunate that these glitches led to a genuine planetary discovery.

While in retirement, Tombaugh maintained an active presence in the American astronomical community. Though no longer formally involved in telescopic investigation, he was most eager to answer questions about his discovery of the ninth planet. He especially enjoyed responding to inquiries about Pluto from young children.

On his 90th birthday, Tombaugh received salutations from all over the world, praising him for his lifetime of work. I spoke to Professor Tombaugh shortly after this occasion, and asked him to relate what it was like working at Lowell Observatory almost 70 years earlier. In his quick raspy voice with its thick prairie accent, he recalled the monotonous labor that paved the way for his discovery:

"It was hard work and very exacting. It was cold. It

was not altogether the most pleasant job to do—and then the scanning of the plates was very tedious.

"The discovery of a planet doesn't happen overnight. It is impossible to find a planet that way. You have to take a photograph, and then, a few nights later, another photograph of the same region, and put the two together. The only clue you have is the motion, and that's not easy."[1]

I asked him how he handled the tedium of so many hours of grueling effort.

"Well it was something interesting to do, and I wanted to do a good job of it, so I stayed with it—and I made it."[2]

Yes indeed, Professor Tombaugh had truly "made it." With regard to the Solar System, he represented the last of the great planet hunters. Now, to stake their claim to alien worlds, planet hunters must look for evidence around distant suns. This is a far more difficult task, one that Tombaugh told me he was glad that he didn't have to face.

On January 17, 1997, Clyde Tombaugh passed away. The world mourned the loss of one of the greatest observational astronomers of the 20th century. The Tombaugh Scholars program, which he established at New Mexico State University, carries on his work by nurturing a new generation of astronomical researchers, eager to scope out new objects in space. The spirit of the discoverer of Pluto pervades the quest for novel bodies far beyond the ninth planet's orbit.

Chapter 3
THE GREAT
WOBBLE HUNT

The heavens themselves, the planets, and this centre,
Observe degree, priority, and place,
Insisture, course, proportion, form,
Office, and custom, in all line of order.

WILLIAM SHAKESPEARE, *Troilus and Cressida*

A Matter of Distance

The ancients believed planets and stars alike occupy similar distances from Earth. Indeed, if one glances up at the
night sky, planets such as Venus and Mars seem to be just
about as far away as the stars that form the Big Dipper.
The lights in the heavens all appear to speckle the same
inky celestial dome. Nothing seems to indicate that any of
these objects is any farther than any other.

Yet we know as a matter of scientific fact that even
the closest stars in the sky are much more distant than the
farthest planets in the Solar System. Proxima Centauri,

the nearest star to the Sun, is almost 10,000 times farther away from us than Neptune and Pluto. While light signals take more than four years to reach us from our closest stellar neighbor, they traverse the entire Solar System in only a matter of hours. And Proxima Centauri is much, much closer to us than the average star upon which we might happen to glance.

As scientists continue to hunt for evidence of planets and planetary life near alien stars, such distance information is of critical importance. Because the nearer a star, the greater the chance of observing planets around it, knowledge of stars' proximities helps scientists know which systems are close enough to be of interest. Close for a star, that is, but not for a member of the Solar System. Fortunately, most of the stars in our part of space have well-known distances based on astronomical measurements.

By looking up at the sky, it isn't immediately obvious that the stars are so far away. How then have astronomers established their great distances? For the stars that are relatively close to us, the answer lies in the method of parallax. Through this technique, astronomers have measured the proximity of nearby stars such as Proxima Centauri.

The parallax method is a natural means of ascertaining distances. In fact, it is one of the ways that the human brain uses to judge how far away an object is. It involves noting the apparent shift in location of a body, relative to its background, when viewed from two different vantage points. The farther an object is from an observer, the greater its shift. By measuring a body's parallactic change in position one can calculate its actual distance.

In a well-known optical effect, the brain uses parallax in determining the field depth of an object. In order to ascertain an object's distance, the brain compares each eye's image of its location. The brain then decides how much of an apparent shift there is between each eye's perceptions. This comparison reveals if the object is in the foreground, background, or somewhere in between. If each eye receives a different image, shifted one way or the other, then the brain determines that the object is nearby. On the other hand, if both eyes perceive the same image, then the brain assumes that the object is far away.

One way of understanding this phenomenon is to perform a simple experiment. Hold a small household object—a spoon, let's say—about an inch in front of your nose. Stare at it with both eyes. Then, alternate opening and closing each eye; when one eye is open, the other should be closed. If this is done correctly, the spoon's image will shift back and forth: left, then right, then left again, over and over.

Now, hold the spoon farther away—at arm's length, say. Repeat the process of looking at it with each eye in succession—first the left eye, then the right eye, then the left again. In that case, the corresponding shift in the spoon's image will be much less pronounced. Clearly, parallax diminishes with distance.

If one were to try the same experiment by blinking one's eyes while gazing at a star, one would soon get disappointed. Stars are so far away that the shifting effect in that case would be visually imperceptible. Then how

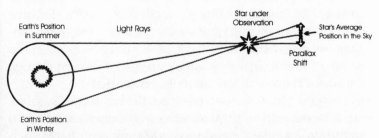

Figure 16. The method of parallax can sometimes be used to determine the distance of a star. This technique relies on the fact that the apparent positions of some stars shift noticeably in the sky from season to season by amounts related to how far away they are.

do astronomers apply parallactic methods in assessing stellar distances?

The answer lies in a clever exploitation of our planet's motion through space. To determine the proximity of stars by use of parallax, researchers take advantage of the fact that Earth's annual revolution around the Sun offers seasonally distinct viewing platforms. In winter, Earth's vantage point is measurably different from that in summer; these positions span almost 200 million miles. Therefore, telescopes surveying the skies during opposite seasons effectively serve as "eyes" hundreds of millions of miles apart. With this broad range of observation points, parallactic methods can be applied to objects trillions of miles away (out to about 100 light-years).

In 1838, parallax was first used to measure an interstellar distance. German astronomer–mathematician Friedrich Bessel noted that a star called 61 Cygni seemed to move especially quickly across the sky relative to the stellar background. This large proper motion (movement compared to a fixed reference) seemed to suggest that 61 Cygni was comparatively close, in the same way a bird whizzing by one's head seems to be moving faster than if it were soaring through the sky. Because 61 Cygni appeared to be so near, Bessel thought it well suited to have its distance found through parallax.

In his experiment, Bessel first observed the position of 61 Cygni, relative to its nearest neighbors in the sky. Using this set of neighbors as a frame of reference, he again recorded the star's location six months later. Then he noted how much its position appeared to shift due to parallax. The parallactic shift he measured was .33 arc seconds; we now know it to be .31 arc seconds. (Arc seconds are units of angular measurement; 60 arc seconds comprise a single arc minute, and 60 arc minutes form a single degree.) Then, using basic trigonometry, he calculated the distance to 61 Cygni to be 8.3 light-years.

The technique of parallax lends itself to an alternative yardstick for stellar distances. If a star's parallactic shift is 1 arc second, then we know it to be 3.26 light-years away. If the shift is half that amount, then we know it to be twice as far away. Taking advantage of this simple inverse proportionality, astronomers often refer to units called parsecs. One parsec—3.26 light years—is the distance equivalent to a shift of 1 arc second.

In 1886, Charles Pritchard pioneered the photographic method of determining parallaxes. He demonstrated that by exposing two different photographic plates and comparing them, anyone could measure distances with unprecedented accuracy. In the beginning of the 20th century, leading astronomers such as Henry Norris Russell, of Princeton University, and Frank Schlesinger, of the Yerkes Observatory, helped to refine this technique further.

Over the years, precise parallactic methods have yielded hundreds of stellar distance measurements. Some of the stars whose distances have been assessed through parallax include Proxima Centauri, Alpha Centauri A and B (each about four light-years away), Barnard's Star (six light-years away), and Sirius (almost nine light-years away). These are some of the closest stars to Earth, ones for which the parallax approach has worked best.

Bright Lights

From distance measurements and other telescopic readings, scientists have gleaned additional information about our stellar neighbors. First of all, the visible stars in our region have been classified according to their apparent and intrinsic brightness. The term "apparent brightness" refers to how luminous a star appears in the sky. Does the star appear as a shining beacon, glimmering for all the world to see, or as a faint hazy dot, barely noticeable?

Sirius, for example, falls into the former category. Also known as the Dog Star, because it resides in the Canis Major (Big Dog) formation that tags after Orion, it has the greatest apparent brightness of any star in our immediate region of space. In fact it is the brightest star in the sky (aside, of course, from the Sun), easily spotted by the casual stargazer on a clear night.

On the other hand, Proxima Centauri is close but faint. It is a small star and puts out little light. Only a trained telescopic observer would be likely to locate it. Therefore, its apparent brightness is low.

A star's apparent brightness alone tells us little about its physical makeup. A more important indicator is its intrinsic brightness. Intrinsic brightness, the actual light output of a star, is a property that can be deduced from its absolute brightness, assuming the star's distance from Earth is known as well. The higher the star's apparent brightness, and the farther the star is away from Earth, the greater its intrinsic brightness. Conversely, the dimmer a star appears, and the closer it is found to be, the smaller its intrinsic brightness.

Intrinsic brightness provides a direct measure of how much power a star actually puts out. This physical parameter helps researchers gauge the strength of a star's nuclear furnace, as well as its capacity to disperse heat from its interior to its surface, and then to surrounding space. And through their knowledge of stellar development, the intrinsic brightness also aids them in assessing its age. Stars tend to shine at different rates during the various phases of their lives.

Another important classification of stars pertains to the type and distribution of their spectral lines. Spectral lines represent the frequency patterns of radiation emitted by an object (or, in the case of absorption spectra, the radiation absorbed by an object). To view these patterns, the light being analyzed must pass through a prism, a diffraction grating, or another optical device that breaks it down into its frequency components. Astrophysicists analyzing stellar light typically find a broad range of constituent frequencies. These consist of both visible and invisible components—with the visible bands appearing as a rainbow of colors, and the invisible bands seen only indirectly.

Stars' spectra tell us much about their chemistries. Each frequency component of stellar light represents the energy produced during a particular atomic transition. Atomic transitions are events in which electrons (negative subatomic particles) within the atoms that make up a star spontaneously lose energy. As an electron drops from one atomic energy level to another, it gives off radiation of a certain frequency. The greater the plummet, the higher the frequency of the light produced.

The specific mechanics of these drops is governed by the laws of quantum physics—a subject beyond the scope of this book. Briefly, these principles state that when an electron drops in energy it must emit a quantum (fixed quantity) of light, released in the form of a photon (light particle). The frequency of the photon is proportional to the amount of energy released by the electron in its fall from a more energetic state to a less energetic state.

This process is analogous to the differing amounts of hydroelectric power produced from waterfalls of various heights. An extremely high waterfall driving a turbine would tend to generate more energy per hour than would an especially low one. Similarly, atomic processes in which electrons drop farther produce more energy—corresponding to higher frequency light output—than those in which electrons drop less.

An atom's electronic energy levels are wholly determined by its type. Hydrogen atoms, for example, possess different electronic structures than do helium or carbon atoms. Therefore, because stellar light spectra tell us the electronic configurations of stars' atoms, they also reveal stars' internal constituents (hydrogen, helium, etc.). For this reason, spectral lines provide researchers with valuable imprints of stars' chemical composition.

Moreover, based on stellar spectra, astrophysicists are able to calculate yet another property of stars—their surface temperatures. The greater a star's surface temperature, the more its spectral lines are shifted toward higher frequencies. The lower the surface temperature, the more a star's spectral lines are shifted toward smaller frequencies. Because high frequencies correspond to colors such as blue and violet, while low frequencies are seen in hues such as red and orange, hotter stars tend to be bluer and colder stars, generally redder than average.

Stellar Menagerie

Based upon observed stellar properties, such as temperature and luminosity (a quantity directly related to intrinsic brightness), astronomers Ejnar Hertzsprung and Henry Norris Russell independently developed a classification scheme for stars at the beginning of the 20th century. Called the Hertzsprung-Russell (H-R) diagram, this representation groups stars according to how hot and luminous they are. In this graph—which has axes representing temperature and luminosity—hot, bright stars are plotted in the upper lefthand corner, while cold, dim stars appear in the lower righthand corner. Hertzsprung and Russell found that the bulk of stars fall along the graph's diagonal; that is, the hotter they are, the brighter they are. Stars that follow this rule are said to belong to the *Main Sequence*.

The Main Sequence includes a number of different star types, ranging from massive blue giants to tiny red dwarfs. Blue giants burn hot and bright and are consequently very easy to see. In contrast, red dwarfs are quite elusive; they shine cold and dim. The Sun falls in between these two extremes. It is a yellow star, of average size, brightness, and surface temperature.

As Hertzsprung and Russell demonstrated, stars evolve throughout their lifetimes. Over the eons, they metamorphise from one stellar type to another. The Sun, for example, will eventually appear quite different from the way it looks today. That is because its main source of power (hydrogen fusing into helium and releasing energy in the process) is finite. Once this supply is exhausted, the Sun will begin to change in character.

Most stars—the Sun being no exception—spend their healthiest years along the Main Sequence. Main Sequence stars are very stable. They neither grow nor shrink very much and emit energy in a steady fashion. That is why the Sun, with its predictable production of heat and light, sustains life on Earth so well. For this reason, scientists expect that habitable planets would most likely be found near vibrant stars of the Main Sequence.

Stars cannot stay on the Main Sequence forever. In their twilight ages, billions of years after their births, they tend to evolve (or devolve, some might say) into increasingly erratic objects. These aberrant bodies include red giants, white dwarfs, neutron stars, and black holes. We shall now consider the former two of these; the latter two we leave until subsequent chapters.

Red giants and white dwarfs occupy peculiar positions on the H-R diagram. They are all mixed up, as stars go. Either too frigid for dazzling brilliance or too scorching for shy faintness, these bodies defy convention. This abnormality is reflected in their peripheral graphical locations. In the H-R diagram, red giants are located above and to the right of the Main Sequence, indicating a high luminosity but low temperature. White dwarfs, on the other hand, are found below and to the left of the Main Sequence, depicting low brightness and high temperature.

Someday, when its primary source of energy—hydrogen fusing into helium—begins to fail, the Sun will swell and become a red giant. Paradoxically, this swelling will commence when the Sun's inner core exhausts its hydrogen fuel source, lacks enough heat-produced pressure to support itself, and starts to contract. The contraction will generate shock waves, hurling the bulk of the Sun's material farther and farther away from the center. The Sun will grow and grow, until it encompasses the region presently occupied by the inner Solar System. Large, cold, and red, it will rapidly dissipate most of its matter and energy into space. Finally, all of the Sun's outer shell will be lost, leaving only the hot, shrinking inner core. The core will be seen by observers (presumably watching from extrasolar planets) as a white dwarf star.

Thus we can summarize the Sun's future behavior in the following manner. It is now a Main Sequence star, and will be so for a very long time. However, sometime in the distant future it will move off of the Main Sequence and become a swollen bright red star. Losing most of its mass, it will then end its life as a dimly burning white dwarf.

Red and white dwarfs are not the only types of small faint stars. Another kind of stellar body, called brown dwarfs, are even dimmer and colder than red dwarfs. Brown dwarfs are stars that lack enough mass for their hydrogen fuel to ignite in the process of fusion. (The amount of hydrogen needed for fusion is approximately 7 percent of the Sun's mass.) As a consequence, they give off no visible radiation. Most of the light they emit is in the infrared range. Infrared radiation is of lower frequency—and hence of lower energy content—than visible light. Because brown dwarfs are so dim, few have been observed.

Giants and dwarfs are broad categories, encompassing a wide range of objects. Color designations clarify things somewhat, but they still do not distinguish every type of star. To classify stars more specifically, the astronomical community has developed a precise nomenclature based on observed spectral properties. It has designated by alphabetical and numerical headings a number of spectral classes that distinguish stars of various chemical compositions.

There are about a dozen principal spectral types, and many more subclasses. Most stars fall into the spectral categories labeled by the letters *A*, *B*, *F*, *G*, *K*, and *M*. Category *A* is characterized by especially strong hydrogen signals, but no sign of other elements. For type *B* stars, astronomers observe faint lines of helium in addition to hydrogen. Stars in the *F* and *G* categories possess metallic lines, as well as hydrogen and helium signals. Finally, types *K* and *M* are classified as such by the distinct spectral signature of molecular compounds (water, for example).

The Sun is a type *G* star, a fairly common designation. Because the Sun sustains Earth so well, many scientists have surmised that habitable worlds would most likely be found around other stars of its spectral class—or of related category *F*. Consequently, searches for signs of extrasolar planets and extraterrestrial life have traditionally centered on these stellar types.

Sirius Disturbances

Most of the billions of visible stars in the Milky Way closely resemble the Sun. Burning hydrogen as its main source of power, the typical shining star is a solid member of the Main Sequence club. Can we assume, then, that the average star is surrounded by a planetary system?

As it turns out, most known stars are orbited by at least one other object. But that doesn't necessarily mean that each is surrounded by a planet. Often, the orbiting body is a stellar dwarf of some kind. (Technically, members of a stellar system orbit their common center of mass, not each other.)

These diminutive companions are not always seen directly. If a binary system (pair of stars) is reasonably close to Earth, and the smaller of the pair is a white dwarf, there is a good chance that astronomers can observe it telescopically. Otherwise, in the case of many brown dwarf and red dwarf companions, they must infer its existence. To discern it using perturbation theory, they must detect its gravitational effects on the luminous star. Or, if they are lucky, the dwarf will turn up on infrared scans.

In 1844, Friedrich Bessel detected the first known stellar companion, Sirius B, a white dwarf near Sirius. He discovered this tiny object while mapping out the motion of Sirius through the sky. He noticed that Sirius tends to wiggle back and forth as it travels; it follows a zigzag pattern, rather than the expected straight line path. Although no companion could be seen at that time, Bessel took the adventurous step of predicting the existence of a dark body having a strong gravitational influence on Sirius.

In 1862, Bessel's prognostication was splendidly confirmed by astronomer Alvan Graham Clark. While observing Sirius with a new telescope, Clark was puzzled by the appearance of a point of light near the star's image. At first he thought this speck was the result of a defect in the telescope. Gradually, he realized it was the hidden object that Bessel had predicted.

Once astrophysical measurements were taken of the object that came to be called Sirius B, scientists, who had never seen a white dwarf before, were amazed. Sirius B's spectral lines pointed to its immense surface temperature. Furthermore, its strong gravitational effect on the motion of Sirius indicated that it is quite heavy. For its small size and faint appearance, it surely is hot and massive—the true hallmark of a white dwarf star.

In the years since Bessel and Clark, white, red, and brown dwarf companions have been seen—or at least made their presence known through their gravitational influence—throughout the Milky Way. Now that many of these dim bodies have been found, celestial bounty hunters have turned to the more daunting task of detecting the more obscure signals of planets.

Considering the difficulties astronomers often have in resolving small celestial objects near stars, how might they know when they have found a genuine planet, rather than merely another dwarf companion? Naturally, if a detected body were found to give off visible radiation, it certainly could not be a planet. Planets do not shine like stars; they cannot emit visible light. They possess no nuclear furnaces to give off such heat.

On the other hand, if a small object discovered near a star were found to be dark, giving off only invisible infrared radiation (or radiant energy too faint to detect), then there would be two distinct possibilities: the companion could either be a brown dwarf or a planet. Observers would need to be careful, then, in coming to a conclusion about its identity.

After all, brown dwarfs and planets do have much in common. Because both produce mostly infrared light, we know that their surface temperatures are roughly similar (give or take several hundred degrees). Moreover, at least some planets (such as the Solar System's gas giants) share with brown dwarfs a composition of mainly hydrogen and helium. Finally, the mass ranges of brown dwarfs and planets appear to have some overlap. In theory, brown dwarfs could exist that are lighter than the heaviest known planets.

In terms of its mass and composition, Jupiter could very well have been a brown dwarf. Like brown dwarfs, the giant planet is almost massive enough for a hydrogen fusion reaction to take place. An alien onlooker viewing Jupiter and the Sun from deep space could very well assume that the former is an unignited star and mistake the two of them for a binary pair.

Their many commonalities may cause one to wonder what distinguishes brown dwarfs from planets. In our quest for alien worlds, how might we tell the difference between "failed stars" and larger members of planetary systems similar to our own? This distinction is important; only in the latter case might we hope for habitability.

Current understanding indicates the major difference between brown dwarfs and planets lies in their origins. Brown dwarfs are thought to have emerged independently, evolving from clumps of material that failed to form into true stars. If the Sun, for example, had started out with only 5 percent of its current mass, it would have become a brown dwarf. Its nuclear processes never would have kicked in, and it never would have begun to shine. It would have ended up as a small, dark, and solitary entity—a glitch on an infrared mapping of space.

In contrast, ordinary planets are believed to have originated as byproducts of the formation of healthy stars. They arose, along with their central stars, from gaseous whirlpools called protoplanetary disks. Over time, gravitational attraction caused the disks' hot central matter to form stars and the disks' cooler peripheral material to clump together into planets. This clumping took place in numerous points of collection throughout each disk. Consequently, whenever this process occurred it

Orion Nebula Mosaic
Hubble Space Telescope · WFPC2

PRC95-45a · ST Scl OPO · November 20, 1995 · C. R. O'Dell (Rice University), NASA

Figure 17. Depicted here is a panorama of the Orion Nebula, assembled from images taken by the Hubble space telescope. The glowing gases associated with this nebula are believed to be similar in makeup to those that gave birth to the Solar System billions of years ago. Thus in viewing the Orion Nebula we are looking through a window to our own past. (Courtesy of NASA)

likely produced whole systems of planets, rather than singular bodies. Thus planets, unlike brown dwarfs, are believed to come in sets, surrounding their mother stars like chicks near a hen.

This points to another dissimilarity between planets

and brown dwarfs. Traditionally, planets were thought to reside fairly close to their central suns (but not too close, or else they'd be absorbed), generally moving in near-circular orbits. This proximity and orbital regularity stems from shared origins. In comparison, when brown dwarfs are paired with shining stars, they often move in eccentric (stretched out) orbits a substantial distance away from their mates.

Scientists believe that most stars have planetary systems because of the apparent role that protoplanetary disks play in the process of stellar formation. These planets are thought to range in size from the diminutive dimensions of Pluto to the gargantuan girth of Jupiter. Formation theories suggest that planets cannot be much more disparate than these. In other words, in the absence of evidence to the contrary, our own Solar System is considered to be of typical proportions.

Barnard's Star Search

The modern-day search for extrasolar planets began in earnest in the late 1930s. Inspired by Adams, Le Verrier, and Galle in their prediction and subsequent discovery of Neptune, and Bessel in his detection of Sirius B, planet-hunters sought to use perturbation theory to find invisible companions to stars. They hoped that at least some of the objects that they expected to discover near stars would turn out to be Jupiter-size worlds.

The center of early investigations of perturbative bodies was an unlikely place—Swarthmore, Pennsylvania. Swarthmore is a tony, upper-middle-class suburb of Philadelphia, half an hour by car or train from the city. It hardly possesses the remoteness we have come to expect for observatory locales. Even then, its skies were brightened by the glare of street lamps from nearby communities. It was certainly a far cry from Lowell's Mars Hill in Arizona.

Yet in those days, a number of major observatories were still located in urban areas. The influence of Lowell and others had not yet inspired astronomers to venture as they do now to outposts on distant mountaintops for the clearest skies possible. Indeed the situation of the Royal Astronomical Observatory at Greenwich, a few miles from London, represented the rule rather than the exception.

The 24-inch refracting telescope at the Sproul Observatory of Swarthmore College, installed in 1912, was still considered in the 1930s to be an outstanding instrument for astrometric scans of stars. Astrometry, the science of celestial measurement, involves precise determinations of the positions of stars and other objects in space. If stellar motion is mapped out properly, this information can be used to locate unseen objects through their gravitational influence.

Many discoveries of dwarf companions to visible stars were made at Sproul. In 1943, astronomer Kai A. Strand presented results showing that one of two stars in the binary 61 Cygni system, 11 light-years away, has an unseen mate that is 15 times the mass of Jupiter, and that 70 Ophiuchi, another binary, also has a third invisible component of 12 Jupiter masses. Invisible companions were similarly found around Lalande 21185 and Epsilon Eridani, stars that are 8 and 11 light-years away, respectively. These discoveries fueled hope that Jupiter-size planets would be found as well. (According to Wulff Heintz of Sproul, however, more recent astrometric research has failed to reproduce these findings, and they are now in doubt.)

The undisputed maestro of Sproul's astrometric program, and the father of extrasolar planet-hunting, was star-mapping impresario Peter van de Kamp. Born in

Figure 18. Sproul Observatory, Swarthmore College. (Courtesy of Swarthmore College)

Figure 19. The 24-inch refracting telescope at the Sproul Observatory of Swarthmore College. (Courtesy of Swarthmore College. Photo credit: Walter Holt)

Kampen, Holland, in 1901, he studied astronomy at the University of Utrecht. After receiving a Ph.D. from the University of California at Berkeley, he was appointed to Swarthmore in 1937. Building on the work of K. A. Strand and others at Sproul, he immediately launched an expansion of Swarthmore's search for hidden extrasolar objects. Over a period of nearly four decades, he led a tireless effort to discover new worlds in space.

Not only did van de Kamp conduct Swarthmore's astrometric mission, he also wielded a baton for Swarthmore's orchestra. As a great aficionado of classical and popular music, he was proud to serve as orchestra director from 1944 to 1954. A friend of Ira Gershwin and Peter "P.D.Q. Bach" Schickele, he even composed his own blues numbers, called "C-Shanty" and "Blackout Blues." When van de Kamp celebrated his 70th birthday, Schickele performed an original tune he had written in his honor, entitled "The Easy Goin' P.V.D.K. Ever Lovin' Rag." Even now, decades after his retirement, van de Kamp is remembered at Swarthmore as both a scientific pioneer and musical virtuoso.

The highlight of van de Kamp's career was his seeming discovery of a planetary system around Barnard's star. Although the finding was later proven false, van de Kamp's efforts brought invaluable recognition to the quest for extrasolar planets. His work helped to generate a great public interest in the discovery of new worlds—a hankering for extraterrestrial adventure that will only be fulfilled through scientific exploration.

Van de Kamp set his sights carefully. Just as composers select themes that evoke the greatest response from their audiences, he tried to pick stars that would indicate the greatest visible reaction to the gravitational forces of surrounding planets. For this purpose, he chose to study the celestial body BD +4-3561, more commonly known as Barnard's star.

Figure 20. Peter van de Kamp, 1901–1995, Professor of Astronomy, Swarthmore College. (Courtesy of Swarthmore College)

Barnard's star was discovered in 1916 by the American astronomer Edward E. Barnard. Only 6 light-years away, it is the fourth closest star to our system. It is a minute object, radiating quite faintly. If it weren't so close, it would have been extremely difficult to detect.

Barnard's star's spectral lines indicate it is quite cool.

Its frequency pattern places it in spectral class M (a red dwarf star), not a particularly promising star-type for habitable planets. However, because its mass is low—15 percent that of the Sun—any planets that it did have would significantly affect its motion.

The most prominent observable feature of Barnard's star is its sizable proper motion. It is the greyhound of stars, racing across the sky at an astonishing speed—for a star, that is. In an average human lifetime, it traverses a path equal to half the width of the moon. This may seem slight, but compared to most stars it is a staggering figure. One would have to be Methuselah, living for centuries, to notice significant motion in the typical stellar body. Yet even a 50-year-old amateur star-gazer might say that he has seen Barnard's star shift its position relative to its background.

Van de Kamp felt that the speed and proximity of Barnard's star made it an ideal subject for astrometric investigation. Any path disturbances caused by hidden objects would be easier to detect in such a fast, nearby star compared to a slower, more distant body. He launched a multidecade, intensive photographic study of the motion of Barnard's star, supplementing data from the Sproul telescope with information dating as far back as 1916. He hoped to map out subtle changes in its path that would indicate the presence of planets.

There were a number of technical challenges that van de Kamp needed to address while carrying out his survey. First, there was the problem of time. Van de Kamp realized that a Jupiter-size planet orbiting Barnard's star would likely take many years to complete a single orbit. After all, Jupiter takes 12 years to journey each time around the Sun. One might expect a large planet orbiting another star to take a similarly long amount of time. Two or three revolutions of a planet—enough time to confirm its existence and reduce the possibility of error—would likely take 20 to 30 years. Therefore, van de Kamp understood that he would need to photograph Barnard's star for at least that interval.

When van de Kamp realized that his program would involve scrutinizing photographic plates for dozens of years, he enlisted the help of an able assistant, Sarah Lee Lippincott. Lippincott had the patience and endurance to scan photo after photo, recording reams of critical data about Barnard's star. In some ways their task was more formidable than Tombaugh's. Rather than searching, as he did, for a single point of light (Pluto) in the sky that moved each day, they were looking for subtler effects that would only make their presence known over decades.

In observing the motion of Barnard's star, the Sproul team could hardly attribute every wobble to the unseen influence of a planet. Rather, the star's apparent path is mainly altered due to parallax. Every 12 months, Barnard's star seems to shift back and forth in its path, due to Earth's changing vantage point as it moves around the Sun. These parallactic shifts are superimposed upon the star's swift pace across the sky. Thus, even without the disturbances of planets, the motion of Barnard's star is a zigzag.

Van de Kamp and Lippincott knew that if they wanted to find evidence of planets near Barnard's star, they would first have to subtract out the pronounced effects of parallax. Once they did so, they could sift through the data more carefully for the delicate imprint of planet-induced disturbances.

To study these subtle path disturbances, they needed a frame of reference. Just as the flutter of butterfly wings is most observable when the insect is perched, minute changes in the motion of Barnard's star could best be seen against a background of fixed points in the sky. For these points, the team chose three distant reference stars, selected for their low apparent motion and minimal parallax.

Between 1937 and 1975, van de Kamp and his coworkers took thousands of photographs with the Sproul telescope, filling two large filing cabinets in the basement of the observatory. Lippincott, ever the loyal assistant, did her part by measuring the great majority of the plates. Van de Kamp then plotted the data, accounting for parallax and other extraneous effects and tested for indications of planets.

By 1963, about a quarter of a century after he began his survey, and more than a decade before he was to complete it, van de Kamp felt he had enough evidence of a planet around Barnard's star to make an announcement. He reported to the American Astronomical Society that he had found an object 1.5 times the size of Jupiter, revolving around the stellar system's center of mass every 24 years. (This was later amended to a period of 12 years.)

The public was gleeful. Ever since Lowell's Martian odyssey turned out to be a wild goose chase, the popular yearning for life on other planets remained woefully unfulfilled. But with the Swarthmore professor's statement, those wishing for evidence of an alternative Earth were able to dream away again.

After researchers in several other observatories, including the respected Allegheny Observatory in Pittsburgh, published papers confirming the Sproul team's results, and even speculated about additional planets, van de Kamp's discovery seemed headed for the history books. Van de Kamp, satisfied after decades of painstaking work, took his astronomical and musical notes and headed for peaceful retirement in his beloved Netherlands. He died there in 1995, at the age of 93.

The Skeptic of Sproul

Why isn't van de Kamp venerated today as the first to have discovered an extrasolar planet? The reason is

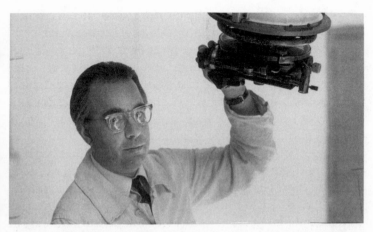

Figure 21. Wulff Heintz (b. 1930), Professor of Astronomy, Swarthmore College. (Courtesy of Swarthmore College)

simple; his results were shown to be mistaken. His data was found to have been falsely interpreted; there is no planet around Barnard's star. And, ironically, the man, Wulff Heintz, who had the sorry task of pointing out van de Kamp's experimental errors, was appointed by him and works in his observatory today.

Heintz was born in the scenic town of Wuerzburg, Germany. In 1953, he received his Ph.D. in astronomy from the University of Munich and was appointed to a teaching position. Fourteen years later, van de Kamp, seeing him as a rising star in his field, appointed him to his current position at Swarthmore. He now shares directorship of Sproul with John Gaustad.

I visited Heintz at Sproul on a sunny April afternoon. I had a strange feeling walking through its doors. Though it was bright outside, the observatory seemed uncomfortable with the radiance of the day. Its dim interior hung on to the gloom of night like a bat in a cave.

Heintz's office, the former darkroom, is a veritable museum of stargazing. Its desks and cabinets are cluttered with logbooks listing every star he has ever observed. He has tracked them all—binary systems, red dwarfs, white dwarfs—stellar bodies of every size, color, brightness, and spectral class. He believes he holds the record for newly discovered double stars: 900 in total.

If Heintz's scientific goal is to find and map novel objects in space, his passion is to check out the claims of others and argue against them if he doesn't believe that they are true. He is particularly dubious about claims of newly found planetary systems. He realizes his skepticism about planets runs counter to the public's interest in alien worlds. But he does not hesitate to take an unpopular position that he considers correct—even if it potentially means being castigated by his colleagues.

"You know the story of Giordano Bruno?" he asked me. "At that time you got burned at the stake if you believed in multiple planets. In the present day U.S. you get burned if you don't believe in multiple planets."[1]

It's not that Heintz rules out the existence of extra-solar planets. It is possible, he thinks, that astronomers will someday find habitable worlds beyond Earth. However, he believes that the media acts too quickly to lend hype to alleged discoveries before these reports are either confirmed or rejected conclusively. Each supposed finding, he says, causes the "news media to report little green men with pink stripes sending UFOs to us."[2] For this reason Heintz feels that he has to "sound a bit more negative to keep things in balance."[3]

Heintz's crowning moment as an astronomical whistle blower (the role, he says, he is often assigned) came in the 1970s, when he realized that something was wrong with the Barnard's star results from Sproul. He noted numerous inconsistencies between the data collected in the 1940s and 1950s. (These problems were pointed out by other scientists as far back as 1958.) Sure enough, he and his colleagues found that something had changed between those decades. In 1949, the telescope had been taken apart, cleaned and reassembled. Somehow in this process, he thinks that one of its lenses may have been rotated with respect to another. This may have led to optical errors.

These misgivings were confirmed in 1973 by Sproul visiting scientist John Hershey. Hershey showed the instruments' optical problems caused similar wobbles in the motion of 12 other stars. This case serious doubt on van de Kamp's data.

That is not all that Heintz believes went amiss. He asserts that van de Kamp overexposed many of the photographic plates. When a plate is overexposed, stellar light tends to become smeared out over a certain range. Consequently, it becomes much harder to pin down the position of a star, and systematic error ensues. Heintz feels these errors weren't adequately addressed.

The best way to correct for these problems, according to Heintz, would have been to monitor the plates' exposures more carefully. In addition, he believes that van de Kamp should have used more reference stars as a background comparison. Heintz tried to persuade van de Kamp to take these measures, but he failed to make an impression on the Dutch astronomer.

"When I told him something was wrong he didn't believe it," Heintz relates.[4]

Later Heintz tried to reproduce van de Kamp's work, taking precautions against overexposure, but found no planets around Barnard's star. Other observers, including George Gatewood of Allegheny Observatory and Robert S. Harrington of the United States Naval Observatory similarly failed to find evidence there of a planetary system. Even Sarah Lippincott, would-be co-discoverer of the "planets," now believes the original results were in error.

"Maybe we exploited the limits of the telescope too far in that one," she said when interviewed.[5]

Today, the story of the planets around Barnard's star has joined the legend of canals on Mars and the tale of Planet X beyond Pluto on the list of well-conceived but ultimately ill-fated astronomical misadventures. Yet we should not be discouraged. For every failed effort in astronomy, there have been thousands of gleaming successes. Even voices of caution, such as Heintz, are optimistic about it future.

Wobble Watchers

George Gatewood and Robert S. Harrington began their independent hunts for extrasolar planets at an uneasy time. During the 1970s, the Barnard's star debacle tainted the field of planet searching—especially with the use of astrometric methods. The wobble approach seemed hopelessly inaccurate and unable to live up to its initial promise.

Ironically, Gatewood and Harrington each played a role in generating this cynicism. Gatewood's 1972 doctoral dissertation from the University of Pittsburgh, based on his research using the 30-inch Thaw memorial telescope at Allegheny Observatory, ruled out the presence of large planets around Barnard's star. (Small planets might exist, but, if so, their wobbles could not be detected.) Harrington's astrometric studies similarly failed to find van de Kamp's planets. These negative results, along with the findings of Hershey and Heintz, called for a major rethinking of how astronomers would carry out the task of looking for planets.

Figure 22. Professor George Gatewood of the Allegheny Observatory. (Courtesy of Allegheny Observatory)

Many scientists bolted from consideration of astrometry and sought alternative approaches for seeking planets. Soon, spectroscopy became the favored means of planet hunting. Yet Gatewood and Harrington, fully aware of their craft's limitations, decided to stick with the wobble approach. In Gatewood's case in particular, not only did he continue the astrometric technique, he also substantially enhanced it.

Fresh out of graduate school, Gatewood was full of enthusiasm about the quest for extrasolar planets. He realized that to increase his chances of success, he would need to use methods of greater precision than those that had been previously employed. In response to a suggestion by Frank Drake that it is easier to measure time than position, he decided to try a novel approach for recording stellar movements using durations rather than distances. With this enhancement, Gatewood found he could achieve

results that were 20 times more precise than those produced by older techniques.

Gatewood's innovative method involves the use of a
special grating placed in front of the Thaw telescope. This
grating is marked off in strips that alternatively block and
pass stellar light. Like partially opened venetian blinds,
the set of bands divides incoming radiation into thin
stripes.

To measure the motion of a star, the grating is moved
at a constant speed across the face of the telescope. This
causes the light from the observed star to appear to blink
on and off each time the opaque sections of the grating
pass by. The effect is similar to that of a strobe lamp. A
detector, consisting of a series of photodiodes (light-
sensitive electronic devices) placed on the other side of
the grating, carefully records this flickering glow.

Once Gatewood has collected enough data within his
detector, he begins his analysis. By measuring the time it
takes for each of the photodiodes to register successive
signals, he is able to compute the path that the star under
consideration is taking. He can see how long it requires to
travel from place to place. In particular, he can determine
its motion relative to a fixed reference frame, consisting of
other stars in its vicinity.

For over a decade, Gatewood has been observing the
motions of a set of about 20 stars, searching for minute
twitches potentially caused by the tugs of planets. By
studying each of these objects at least 25 times per year he
has been able to pin down its movements within an accuracy of 1 milliarcsecond (about one four-millionth of one
degree). Gatewood claims his instruments and techniques are powerful enough to detect any planetary system located less than 40 light-years away. After many
years with nothing definitive to report, he finally announced in 1996 the discovery of a planet with a mass
close to that of Jupiter around the star Lalande 21185.

(As of this writing, Gatewood's finding has not yet been confirmed.)

Robert Harrington is probably remembered best for his 1978 co-discovery, along with James Christy of Charon, the only known satellite of Pluto. Yet he was a remarkable extrasolar planet-hunter as well. His meticulous approach won him the admiration of his colleagues. When he died suddenly in 1993, his passing was mourned by the entire astronomical community.

Harrington's astrometric method was more conventional than Gatewood's. Instead of using a special apparatus, he employed traditional parallax techniques to ascertain the motions of stars. Nevertheless, he was quite efficient in doing so. Together with his colleagues at the United States Naval Observatory, he observed in his lifetime roughly 1000 stars, probing for distinctive signs of new worlds.

In 1984, Harrington reported evidence of an extrasolar planet around the red dwarf Van Biersbroeck 8. After the newly found body (dubbed VB8-b) was measured through its gravitational influence on its central star, its mass was estimated to be about 50 times that of Jupiter. Scientists immediately realized that the object was too massive to be a planet and was probably a brown dwarf.

Excitement grew considerably when Donald McCarthy and coworkers at the University of Arizona spotted VB8-b independently, using a spectroscopic approach. Ignoring the fact that the body probably wasn't a planet, the media began to report the first extrasolar world had been found. News programs on all three networks heralded the finding as a major discovery.

A few years after VB8-b was first detected, several research groups tried to conduct an intensive study of its properties. To their bewilderment, the object could not be found at all. A comprehensive search was carried out, but nothing was seen. Today, astronomers generally believe

VB8-b doesn't really exist. Like van de Kamp's planets, its imprint was likely the product of experimental error. Curiously, the news media that trumpeted the initial discovery never reported later evidence of VB8-b's nonexistence.

Harrington did not live to complete his quest for extrasolar planets. With his untimely death, astronomic research at the United States Naval Observatory was dealt a harsh blow. Nevertheless, his colleagues and students have continued his work. The great astronomical wobble watch carries on.

Astrometry is hardly the same discipline, however, that it was in the time of van de Kamp, or even in the early days of Gatewood and Harrington's research. In the past few years, there have been revolutionary changes in the field. Photographic plates are on their way out, gradually being replaced by light-sensitive electronics, such as charge-coupled diodes (CCDs). Tiny light-sensitive silicon wafers, each barely one-sixth of one inch across, CCDs work like miniature electronic cameras. Placed at the focal points of telescopes, they sense the light of single photons (light particles) and convert that energy into electric signals. These pulses are transmitted to computers, where they can be displayed on monitors. Comprising much more precise means of detection than conventional photography, a single CCD placed at one end of Yankee Stadium on a dark night could easily sense a firefly sitting at the other end.

The other major transformation taking place in astrometry pertains to a novel type of telescopic system known as adaptive optics. One problem with traditional telescopes is the distortion of starlight as it passes through heat zones in Earth's atmosphere. This twinkle effect represents a substantial source of error in measurements of stellar positions. Although this problem can be ameliorated by conducting searches in outer space with devices such as the Hubble space telescope, the Hubble is not designed for astrometry.

Adaptive optics, hooked up to ground-based telescopes, constitute a promising, cheaper alternative to time on expensive space-based instruments. With these specially designed optical systems, the twinkle effect can be reduced considerably. Adaptive optics first analyze the light coming from a star to determine the effects of atmospheric turbulence. Within milliseconds, a computer program determines how best to adjust the telescope's mirrors in order to correct for any distortion. The mirrors are then tilted by a minuscule amount, just enough to diminish the twinkling. This process repeats as needed to produce the clearest images possible.

The adaptive optics group at the Steward Observatory of the University of Arizona have been pioneering this approach with their multiple mirror telescope. They have built an experimental system that aims a bright laser guide wherever the telescope points. After measuring the effects of atmospheric distortion on the laser image, a special mechanism adjusts the telescope's mirrors accordingly. The reflectors used have numerous moveable components—ideal for small adjustments. For their wonderful flexibility, reflective surfaces of this type are nicknamed "rubber mirrors."

The Steward team's studies look promising. They have found that their method generates much sharper pictures than could be produced through conventional means. Proud of their innovations, they have been actively exporting their system to other observatories.

In the past decade, the astrometric search for extrasolar planets has been eclipsed by newer spectroscopic methods. Because of several well-publicized successes in finding planetary evidence by use of stellar Doppler shifts (a method discussed in Chapter Five), the wobble approach has seemed consigned to astronomy's back burner. Perhaps with its traditional reliance on ancient telescopes, manual instruments, and photographic plates, astrometry now appears rather frayed, like used film left on a darkroom floor for ages. Researchers hope, however, that the advent of adaptive optics and CCD detectors will soon pump some life back into the old field. With these new technologies, astrometry will likely regain its role as an important planet-hunting technique.

Chapter 4
RIDDLE OF THE PULSAR PLANETS

The proof that objects of planetary size do exist outside the Solar System indicates that our planets are not unique and uncommon anymore.

ALEXANDER WOLSZCZAN, *Penn State astronomer, 1995*

Duelling Announcements

In the 1490s, great sailing ships bore intrepid European explorers across vast, treacherous seas to mysterious new lands. Those who embarked on these voyages chanced to venture beyond the realm of familiar territories without the use of adequate maps. As they entered the unknown, they trusted in the guidance of divine providence to steer their vessels to terra firma.

Interplanetary explorers heading across the far more formidable ocean of space (once the technology to do so becomes available) would not want to take such risks. If they were to sail off into the void they would certainly require detailed sky atlases, with planetary targets carefully pinpointed. Otherwise, with space being so vast, failure to find such worlds would be almost inevitable. That is why an age of interplanetary voyages would require a period of planetary cartography: the meticulous mapping out of where possible worlds might lie.

Five hundred years after Columbus sailed to America, paving the way for the great terrestrial Age of Exploration, the first stirrings of an Age of Planetary Discovery have appeared. This new era has been launched by another enterprising European, Polish astronomer Alexander Wolszczan. In the early 1990s, he announced his finding and confirmation of the first known planets beyond the Solar System. By discovering these worlds, he has placed his signature on the first page of a new atlas of the cosmos—the requisite guide for future space exploration.

The circumstances of Wolszczan's extraordinary announcement were strange indeed, befitting such unusual worlds. His first public statement was made at the January 1992 annual meeting of the American Astronomical Society in Atlanta. In a bizarre twist of fate, he was preceded on the podium by another astronomer, Andrew Lyne of the University of Manchester, England, who had come to retract a similar discovery. A planet Lyne had previously claimed to have found really didn't exist. Thus, at the same time Wolzsczan's planets flickered into view, Lyne's blinked out of reality.

To fathom the irony of such circumstances, imagine if at the same time Columbus reported his discovery of the Americas, another sea explorer announced to the Spanish royals the finding of lost Atlantis. Then, consider the blow to the credibility of Columbus if the second captain's report turned out to be the observation of a mirage. Perhaps King Ferdinand and Queen Isabela of Spain would

Figure 23. Alexander Wolszczan (b. 1946), discoverer of pulsar planets. (Courtesy of Pennsylvania State University. Photo credit: Greg Grieco)

have discounted Columbus' discovery, and exploration of the Atlantic would have been dealt a critical setback.

Fortunately Lyne's error did not have such an effect on the search for planets. Future discoveries (such as Wolszczan's) were assigned credibility nevertheless. Perhaps Lyne's results even whetted the appetite of the astronomical community for future findings and set standards by which they could be judged.

Certainly Lyne's straightforward approach served to help matters. It is enlightening to examine how he handled what amounted to an embarrassing, and potentially career-damaging, situation—the announcement of a new planet turned mirage. It is an example of scientific integrity at its finest.

Lyne's retraction at the Atlanta conference was an act of sheer bravery. Less than a year before the meeting, he had been lauded by the international press for the apparent first sighting of an extrasolar planet. He was all set to present additional confirmation of his earlier results but now had to make an embarrassing admission of error.

Beginning with the statement, "this talk is not the one I was originally proposing to give,"[1] Lyne nervously explained to the shocked audience what had happened. The audience listened intently to his somber account of a planet that almost was.

The Pulsar Planet that Almost Was

It had all started out very promising. Using radio telescopic techniques, Lyne and his colleagues Matthew Bailes and Setnam Shemar were conducting a five year survey of 40 pulsars at the Jodrell Bank Observatory, when they discovered that one of them behaved as if it were affected by an unseen body. The presence of a planet around the pulsar seemed to them to be the most logical explanation.

Recall that pulsars, the rapidly spinning, ultradense cores of massive extinct stars, emit detectable radio waves as they rotate. The fluctuations in the strength of these signals can be monitored by astronomers to provide insight into the properties of a pulsar under study. In particular, they can be used to assess the pulsar's rate of rotation, as well as its distance from Earth.

Generally speaking, pulsar signals vary in an extremely regular manner, like the ticking of an extraordinarily precise clock. Each time a pulsar spins, it drags its focused beam of radiation around a tight circle in a manner reminiscent of a lighthouse beacon. To observers on Earth, these radio wave emissions appear to increase and decrease in strength periodically. The time for a single cycle of variations to occur—an interval known as the period—ranges for most pulsars from one half-second to one second.

To ascertain the location of a pulsar, astronomers normally use a radio telescope to measure the intensity and direction of the incoming radiation. These quantities are then compared with those that would be expected for a pulsar of a given position. Once this comparison yields the pulsar's location, a computer program is employed to predict its future movements. Ultimately, this information is used to track the pulsar across the sky.

Assuming that a pulsar's position is known, researchers can calculate the expected time for its signals to reach Earth. This astronomical measure, called the pulse arrival rate, depends on Earth's location, as well as the pulsar's. Because both Earth and a given pulsar execute complex motions through space, a detailed formula must be used to determine this quantity. Factors such as Earth's rotation, its revolution around the Sun, and the Sun's motion through the galaxy must all be taken into account. With such a complicated calculation, there is substantial room for error, as Lyne was to find out.

Lyne and his colleagues had detected small periodic variations in the pulse arrival rate from the pulsar PSR 1829-10, located near the center of the Milky Way. The time for its signals to reach Earth seemed to alternate—first increasing, then decreasing—over intervals of six months. Some signals seemed to arrive early; others, to come late. From a five year study of these fluctuations, using the 250-foot diameter Lovell telescope at Jodrell Bank, Lyne had concluded (falsely, it turned out) that this pulsar was orbited by a planet the size of Uranus. The planet's gravitational pull, he had surmised, was causing the pulsar to wobble back and forth. In turn, these vibrations were creating cyclical variations in the arrival times of the pulsar's radiation.

This effect is analogous to that of a flickering lantern, hanging from a cord, swaying back and forth in a strong wind. Imagine viewing such a lamp on a dark, foggy night. When the blinking light is blown closer by the wind, its signal might appear more distinct and closer than when the lamp is tugged farther away. At some points, the flickering light might seem brilliant in the nighttime air. At other times, it might appear barely noticeable. In short, the tugging effects of the wind would affect the visible proximity of the lamp.

Lyne thought that the rhythmic changes in arrival times of the blinking light from PSR 1829-10 was similarly due to the tugging action of an external force. In this case he believed it was the gravity of a planet pulling on the pulsar that was causing these variations.

Lyne, Bailes, and Shemar published their findings to much acclaim in the July 1991 issue of *Nature*. As in the case of van de Kamp's and Harrington's "discoveries" years earlier, newspaper headlines heralded the detection of an alien planet without waiting for independent confirmation of the results.

Although the press seemed satisfied that an extrasolar planet had been found, the scientific community required more than just the account of a single team working for a few years reporting an initial result. To prove conclusively that there was a planet orbiting PSR 1829-10, Lyne realized that he needed to take even more measurements and to check and recheck his calculations. After doing this, he planned to present his bolstered conclusions at the 1992 Atlanta meeting.

Less than a week before the conference, Lyne realized that he had inadvertently left one crucial factor out of the equation. He had failed to account for the elliptical nature of the Earth's revolution around the Sun. Because the Earth's orbit is an ellipse, not a circle, the distance between the Earth and the Sun varies periodically. This seasonal variation must be included in any estimation of how the distance between Earth and another body changes over time. Lyne and his research team had neglected to program in this factor. Therefore, the fluctuations he and his coworkers had seen were the result of Earth's relative motion, not the pulsar's. After appropriate corrections were made, the planet simply vanished. Lyne relates that he "froze in horror" as the planet "evaporated."[2]

Once again, another planetary discovery turned out to be a mirage. The planet around PSR 1829-10 joined van de Kamp's substellar objects and VB8-b as another in a string of illusory bodies—shadow worlds produced by experimental error.

Many researchers would have flinched at admitting such a grave mistake. But Lyne and his colleagues immediately issued a statement to *Nature* explaining what had happened. (The article appeared in print the day after the conference.) After submitting his written retraction, Lyne was faced with the painful act of making his scheduled presentation at the meeting.

The audience responded to Lyne's confession with a long ovation. "We applauded him for his honesty," said Stan Woosley, a University of California at Santa Cruz astronomer who was present at the meeting. "He is a very careful scientist. He recognized his error and reported it promptly."[3]

Lyne remains undaunted and energetic in spite of his mistaken endeavors. At the Jodrell Bank Observatory in England, as well as at the Parkes Radiotelescope in Australia, he continues to study the varied phenomena of pulsars and hopes to find solid examples of pulsar companions.

The Mystery of Pulsars

Even before Lyne issued his retraction, a number of astronomers were quite incredulous that planets could be found around ordinary pulsars. This skepticism stemmed from the then-understanding of the origins of pulsars. The scientific picture of how pulsars evolve in the turbulent centers of supernova blasts did not seem to allow for the possibility of planets.

Pulsars are thought to originate as end-products of the death of very massive stars—those with much more material than the Sun. As we have discussed, when the Sun (or a similar star) dies, it will pass through a red giant phase, and eventually become a white dwarf. The dwarf star will shine dimly for a while and then eventually burn out. On the other hand, if a star starts out more than twice as massive as the Sun, its end game will be played out quite differently. Its demise will be far more eventful than that of a solar mass star.

The death struggle of a very massive stellar object begins when the star exhausts its available hydrogen fuel. It can no longer shine through the usual process of fusing hydrogen nuclei together into deuterium and helium nuclei. In the absence of a readily ignitable fuel source, the inner core of the star consequently shrinks down under its own weight. Meanwhile, the outer gaseous envelope of the star, no longer held tightly inward by the core, starts to expand outward. Eventually the star becomes a colossal ball of gasses surrounding a compact core—a mammoth body known as a supergiant.

As the core of a supergiant contracts, it becomes denser and denser, growing hotter and hotter. Soon it is fiery enough to fuse together light nuclei into heavy elements. Over time, the process of nuclear fusion taking place in the core transmutes simpler materials into more and more complex substances. Like snap-together building blocks combined, step by step, into increasingly intricate designs, basic elements are gradually consolidated into heavier nuclei. With each element fueling the production of the next in the chain, carbon, oxygen, and other massive substances are gradually constructed from simpler materials.

Eventually, the process of nuclear fusion reaches its absolute limit. The element iron, once produced, cannot efficiently fuse together into heavier substances. For it to transmute into more massive elements, it must consume existing energy rather than release more of it. Consequently, once enough iron is built up within its core, the supergiant cannot gain more energy through fusion. The only way that the core can sustain itself energetically is to collapse.

The contraction of a supergiant's core takes place with incredible haste. In a fraction of a second, it collapses into a ball of ultradense material only tens of miles across. The flood of energy released in this sudden implosion disperses rapidly throughout the outer envelope of the star, blasting it off into deep space. In an instant, the supergiant sheds the overwhelming bulk of its volume—an event which appears to observers on Earth as a supernova explosion.

A supernova explosion does for a star what an Academy Award does for an unknown actor. One moment a stellar object might be languishing in obscurity, barely visible except to the most stalwart of astronomers. Perhaps it is not even listed in Boy Scout sky-survey handbooks, the stellar equivalent of *Variety*. Unlike the North Star or Sirius, its name isn't even known by the average park ranger.

Then, suddenly, the star bursts onto the scene in a torrent of high frequency energy. A gush of neutrinos (fast-moving elementary particles) heralds its new-found fame, followed by a wave of photons (light particles). Once a second-rate bit-player, now it is the superstar of the night sky. For a time it is the brightest object in the nocturnal heavens save none. With any luck, assuming there are no political scandals that week, its proud image might very well be featured on the covers of *Time*, *Newsweek*, and *Sky and Telescope*.

Alas, fame is fickle. As the energy from the supernova spreads out into deep space, the explosion site grows dimmer and dimmer. A faint, colorful haze, fashioned from leftover gases, known as a supernova remnant, veils the region where the starburst took place. Soon, the central blast can no longer be visually observed. The remnant provides the only visible trophy of the supernova's brief sojourn upon the celestial stage.

At that point, only the shrunken core is left of the original star. The composition of this relic is quite peculiar. Rather than containing ordinary atoms, formed of protons and electrons, it is composed instead of a substance called neutronium. Neutronium, an extremely dense state of matter, is formed when protons and electrons are crushed together under great pressure to form neutrons. The density of neutronium is so great that a single spoonful would weigh about 100 million tons. Within the stellar relic, it is packed so tightly that it essentially comprises a single atomic nucleus, albeit with trillions and trillions of neutrons.

As this neutronium core is created, it begins to spin more and more rapidly, until it is whirling about its axis at lightning speed. The reason for this steady increase in rotation lies in the principle of the conservation of angular momentum. Angular momentum is a physical quantity that depends on the mass, extent, and rotational speed of a body. In the absence of outside influences it is conserved, meaning that it cannot be created or destroyed. For this reason, when an object contracts, it must spin more quickly. What it loses in breadth it must gain in rotational speed. (This is the same reason why twirling dancers draw their arms inward when they wish to spin faster.)

By that point the core remnant has become the compact, twirling object that astronomers call a pulsar—named for its rapid-fire pulses of radio waves. (Because of their composition, these bodies are also referred to as neutron stars.) Each time a pulsar spins, it radiates energy into space along narrow beams. These periodically generated bursts of radiation travel through space and are picked up by astronomers' radio telescopes as regularly blinking signals.

When radioastronomers, such as Andrew Lyne and Alexander Wolszczan, detect repetitive pulsating signals, they surmise that the source of these radio waves is an extinct star. The regularity and wavelength range of such pulses rule out other astronomical possibilities and point to pulsars as their origin. It has only been in recent decades, however, that scientists have had the experience and insight to come to such conclusions. When pulsar signals first were detected in the 1960s, they were thought to be the broadcasts of intelligent extraterrestrials. Only after researchers delved more carefully did they discover the true nature of the pulses they saw.

In 1967, graduate student Jocelyn Bell, working for Cambridge astronomer Anthony Hewish, was examining data taken from a new radio telescope Hewish had constructed. She expected the data to be the typical jumble of noise, with no apparent pattern. To her astonishment, in sifting through the information, she noticed a steady rhythm of evenly spaced blips. One tick after another, something was sending out signals in clockwork fashion.

She checked the lab, making sure that nothing internal was clicking off these pulses.

When she showed the results to Hewish, he was incredulous at first. Once he confirmed her discovery, they began to wonder what extraterrestrial phenomenon could generate such pulses. Could these strange signals be the welcoming call of an other-worldly civilization? They began seriously to confront the possibility that they were witnessing efforts by another race to contact Earth.

Alas, there were no "little green men" (as they put it) to be found. In researching possible origins of the radio waves, Hewish stumbled upon a paper by renowned Russian physicist Lev Landau, describing the possibility of rapidly spinning collapsed stellar cores. Hewish soon realized that such collapsed cores would generate pulses of energy similar to those detected by his group. He estimated that a pulsar of approximately six miles in diameter, rotating once every one and one-third seconds, would produce the signals observed. For his discovery, he won the 1974 Nobel Prize for physics.

Since the time of Bell and Hewish, pulsar detection has become a veritable industry. With a variety of sky-scanning radio telescopes, astronomers have catalogued thousands of pulsars throughout the cosmos. By examining their rhythms and spectra, researchers have learned how to estimate their masses, distances, and other properties.

Considering what is now known about the origins of pulsars, it is not surprising that Lyne's announcement of his discovery of a pulsar planet baffled many radio astronomers. Assuming that the planet was once in orbit around the star that spawned the pulsar, they wondered how such a body could have survived the fiery collapse of its parent sun. They doubted that there was any chance that a planet could have withstood the impact of a supernova without being hurled off into space. Naturally, the astronomical community wasn't all too surprised when Lyne's discovery turned out to be a mirage.

In a Flash

Alexander Wolszczan, though interested in astronomy from an early age, never imagined that he would be the one to discover—and confirm the existence of—the first known extrasolar planets. Born in 1946, he grew up in Eastern Europe at a time when outer-space fantasies were seen as a pleasant diversion from bitter post-war realities. While newsreels and magazines heralded the launching of the first Sputnik satellites (by the powerful neighboring Soviet Union), popular visionary writers such as Stanislaw Lem crafted futuristic fables of the space age. And, like many youth of his day, young Alex had his gaze fixed steadily upon the stars. Encouraged by his father, he devoured all that he could find written about the cosmos. For his university education, he enrolled in the radio-astronomy program at Nicolas Copernicus University in Torun, Poland, where he completed his Ph.D. in 1975.

In the 1970s, the decade after Bell and Hewish's discovery, the study of pulsars was considered a promising new field. Wolszczan decided to specialize in this area, feeling that it was "still a relatively new topic, full of interesting issues to resolve."[4]

In the years to follow, Wolszczan became an expert in his field, taking measurements with the huge radiotele-scope at Arecibo Observatory. He became adept at using the 1000-foot wide instrument to detect faint signals from distant objects. From the early 1990s onward, as a sub-specialty, he began to explore the novel domain of milli-second pulsars.

Millisecond pulsars form a special class of extremely rapidly spinning celestial objects. With periods measured in the thousandths of a second, they rotate hundreds of times faster than ordinary pulsars. These stellar whirlwinds are thought to be remnants of supernova explosions of binary star system members.

The first millisecond pulsar detected, called PSR 1937+21, was discovered in 1982. Lying in the constellation Vulpecula, it was found to spin more than 600 times per second. Since that time, dozens of other rapidly rotating pulsars have been found.

Scientists have developed detailed scenarios for how millisecond pulsars are created. In these models, a millisecond pulsar is postulated to begin as a supergiant locked in mutual orbit with another star (its binary companion). Suddenly the supergiant bursts, blasting most of its material off into the void. Left behind is a compact neutron star, the remnant of the supergiant's core. This star picks up rotational speed as it contracts, and slowly loses energy, sending off pulsed radio signals.

So far, what we have described might apply equally well to an ordinary pulsar's formation. However, around that point, the similarities start to fade. Soon the existence of a companion star begins to play an important role in the millisecond pulsar's evolution.

The relationship between a pulsar and a living star in a binary system can best be characterized as that of a thief and his victim. The pulsar is so dense, it has an extremely concentrated gravitational field. Imagine the pull of the Earth condensed into a space less than the size of a house. That's the strength of a pulsating neutron star. When another hapless star is nearby, this power can be used by the pulsar—like a cosmic Jack the Ripper—to wrest the life out of its mate.

This scenario plays out as follows. With its mighty gravitational force, the compact object draws its companion closer and closer to itself. At the same time, it gradually bleeds from its partner considerable quantities of matter and energy. In particular, it steals much of the rotational power of the other star. Gradually, the pulsar builds up its rotational speed at the expense of its companion, until the former is whirling around hundreds of times per second, and the latter is barely spinning at all.

In recent years, scientists have detected this process at work. A ravenous stellar body, dubbed the "black widow pulsar," has been observed in the act of gobbling up a nearby star. A steady stream of matter feeding the pulsar has made its presence known through its energy signature. Characteristic signals indicating the infall of material have been picked up from that region. Theorists expect that, as it cannibalizes its partner, the black widow will speed up over time and eventually become a millisecond pulsar. They estimate it will likely reach that state in hundreds of millions of years. Other probable millisecond pulsars in formation have been seen as well, similarly caught in the act of consuming their prey.

Their unusual properties and relative rarity make millisecond pulsars of great interest to astronomers. Observers of these anomalies hope to glean from their behavior critical information about the nature of binary systems, as well as data concerning the unfolding of supernova explosions.

Realizing the merits of studying such objects, Wolszczan embarked in 1990 on an effort to pinpoint more of them. He took advantage of the fact that in the early part of that year the Arecibo radio telescope was closed to outsiders for structural repairs, but still usable by staff members for sky surveys. Exploiting this window of opportunity, he scheduled night after night of observing time and scanned far-flung regions of the galaxy for signs of regular pulses. He soon discovered two new millisecond pulsars, PSR 1257+12 and PSR 1534+12, which he began to examine in detail.

Of the two new pulsars that Wolszczan found, PSR

1257+12 proved to be the most interesting. Located 7,000 trillion miles away in the constellation Virgo, its signals were estimated to have taken approximately 1600 years to reach Earth. Thus, the blinking he observed actually took place almost two millennia ago—around the same time the library of Alexandria in Egypt burned to the ground.

Wolszczan timed the precise arrival rates of pulses from the star and tried to establish a pattern. He expected to observe the clockwork flashes of a regularly signaling pulsar. Instead he was baffled to find that the signals possessed a not-quite-periodic beat. The pulsar's ticking was off, as if it were a defective metronome. Some of the pulses arrived several milliseconds earlier than expected; others, a few milliseconds later. He decided to consult with his colleague Dale Frail of the National Radio Astronomy Observatory in Green Bank, West Virginia, and reanalyze this curious data together.

Second-Hand Worlds

Given the beauty of the sky on a clear night, astronomy would seem to be a glamorous profession, full of glitter and glory. One would think it would be utterly romantic to spend one's career gazing at the stellar lights that shine above us every evening and to rhapsodize in a profound manner about their appearances and motions. Yet, as with most jobs, on a day-to-day basis, the realities of the profession typically turn out to be far more prosaic.

With the help of Frail, Wolszczan spent the summer of 1991 staring at figures—not blazing, cascading comets or brilliant, multihued nebulae, but just tedious, commonplace numbers. By churning through his massive pile of data, Wolszczan wanted to figure out why the rhythm of PSR 1257+12 wasn't regular, why the pulses weren't spaced out in fixed intervals. The answer might be found only through painstaking analysis—by considering, and then either including or ruling out, each reasonable option.

Meanwhile, that same summer, Andrew Lyne and his colleagues were announcing over in England their apparent discovery of a world around the ordinary pulsar PSR 1829-10. As we have discussed, this announcement sent shock waves throughout the international astronomical community. For the first time it was conceivable that planets could be found orbiting extinguished stars. Radio telescopic surveys, it seemed, might be a good way to find long sought extrasolar planetary systems.

Within this context, Wolszczan and Frail checked their data for the possibility that the tugs of planets were causing the pattern of signal delays. Sure enough, they found they could explain their results by assuming that at least two worlds were circling the pulsar. One planet appeared to have an orbital period of 66.6 days and the other 98.2 days.

By supposing PSR 1257+12 has a typical pulsar mass of 1.4 times that of the Sun, Wolszczan and Frail found they could make predictions about the planetary system around the pulsar. They estimated that the inner pulsar planet orbited the star at slightly over one-third of the Earth–Sun distance, and the outer world at almost one-half of the Earth–Sun distance. For comparison's sake, the average radius of Mercury's revolution around the Sun, at about two-fifths of the Earth–Sun distance, falls in between these two values.

Using the same set of suppositions, they also calculated the masses of the two worlds. Each was found to be about three times as massive as the Earth, the inner world somewhat heavier than the outer. These masses were estimated by discerning the gravitational tugs required to have generated the pulse time delays observed—the greater the mass, the stronger the gravitational perturbation, and the lengthier the delay.

Wolszczan is a cautious man. He elected to wait until he had taken many months of continued observations before announcing his discovery. As more and more time passed and the predictions of his model proved accurate, he decided to publish his results and present them at the 1992 meeting of the American Astronomical Society.

All set to state his case for new planets, Wolszczan was dismayed to find the conference abuzz with news of Lyne's admission of error. As scheduled in the meeting's program, Wolszczan was to announce his discovery shortly after Lyne's talk—a retraction of results that superficially appeared similar to his own. Such circumstances had potential for embarrassment, yet Wolszczan went on and his talk was well received.

Some observers at the conference were concerned that Lyne's retraction would damage Wolszczan's credibility. Actually, it had a mixed effect. On the one hand, attendees questioned Wolszczan more carefully about his methods, making sure that he hadn't made the same mistake as Lyne. They sifted through the Penn State astronomer's results with even more rigor than they would have ordinarily. As Wolszczan has confessed, "it wasn't the easiest time of my life!"[5]

On the other hand, Lyne had introduced the subject of pulsar planets to the astronomical community. Before Lyne's work, researchers had scarcely thought about the possible existence of planetary objects near such bizarre bodies of pulsars. Since the topic was already on the table by the time of the conference, Wolszczan had an easier time making the case for his own pulsar worlds.

Furthermore, conference participants realized that the theoretical case for planets around *millisecond* pulsars was far more sound than that for planets around conventional pulsars. They were optimistic that Wolszczan's results, if confirmed, would have a sound basis in astrophysical theory. Therefore, in this regard, there was less skepticism about Wolszczan's findings than there was for Lyne's.

This optimism stemmed from the specifics of millisecond pulsar formation models. According to these scenarios, when a millisecond pulsar is formed, it drains matter and energy from its binary companion. If this process proceeds until it has reached its limit, then in theory the companion becomes completely pulverized. Nothing is left of the original star except interstellar dust and other forms of debris. Due to the influence of gravity, this matter eventually settles into orbit around the pulsar.

What happens next, according to these models, is the creation of a novel planetary system around the pulsar. Over the eons, under the influence of its mutual gravitational attraction, the fine material from the demolished star begins to coalesce into chunks of greater size. Gradually this process leads to the building up of larger and larger pieces, called planetesimals (rocky components of planets). Ultimately the planetesimals fuse together into planets, revolving around the millisecond pulsar.

The bodies created through such processes would be second-hand worlds, the product of cosmic recycling. Unlike primary planetary systems—the Solar System, for example—these worlds would be spawned through the death of a star instead of its birth. Rather than being forged in the hearth of stellar creation, they would be smelt in the pyre of stellar destruction. They would be fossil orbs, not vital spheres.

No wonder scientists were delighted by the discovery of planets around millisecond pulsars. Their existence solves the riddle of what happens to the companions of millisecond pulsars after they are broken apart by the monumental gravitational forces associated with such compact objects. The answer is simple: they form the fertile loam in which new planets can grow.

It has also become clear why planets are not as likely to be found near ordinary pulsating stars. Normal pulsars are either solitary (without a binary companion), or if they are in a binary pair, they are with a companion that is basically intact. In other words, they belong to a younger system and haven't had time, as of yet, to wreak much havoc on another star. We know this because of their rate of rotation. If they had already caused damage, they would have profited by stealing the rotational energy of the companion, and would be spinning faster by now. Hence, in short, pulsars that have generated the destructive force needed to demolish a neighbor and create planets would have also gathered the rotational energy needed to become millisecond pulsars.

As the 1992 meeting in Atlanta drew to a close, participants had much cause to celebrate. It appeared that astronomy's long-standing goal of detecting extrasolar planets had finally been realized. Furthermore, it seemed that observational evidence was lending support to current theoretical models of how pulsars and millisecond pulsars are formed. However, Lyne's unfortunate experience with a planet-turned-ghost lent a whiff of caution to this atmosphere of triumph. Before the banners could be unfurled, it was clear to all attendees that the phenomenon of planets around millisecond pulsars required further validation.

The Confirmation Challenge

After the conference was over, Wolszczan took up the critical task of confirming his own tentative results. To do so, he realized he needed to observe the millisecond pulsar's radio output for a few more years, until his data was unmistakably accurate. Even that would not be enough. Other astronomical effects that could mimic the presence of planets would have to be ruled out. For example, what if the distorting effects of an intervening object were causing the fluctuations? Or perhaps some complex series of disturbances within the pulsar was creating the radio signal effects. Because two planets, rather than one, were apparently detected, these possibilities seemed highly unlikely. It was hard to imagine other natural phenomena duplicating the precise signatures of two orbiting planets. Still, to make a bold claim in science, absolute proof is required. And what could be a bolder claim than the discovery of the first known planetary bodies beyond the solar system?

Fortunately, soon after Wolszczan's report, radiophysicists Frederic A. Rasio, P. D. Nicholson, S. L. Shapiro, and S. A. Teukolsky of Cornell University came up with an unmistakable observational test for the existence of a planetary system around PSR 1257+12. Their test relied on the fact that the planets were seen as orbiting the pulsar in close to a 2:3 ratio. That is, each time the outer planet circles the star twice, the inner planet travels three times around the star. About every 200 days (approximately two revolutions of the outer world and three revolutions of the inner world), the two planets line up in their orbits, making them both on the same side of the pulsar. Therefore, their gravitational effects on the pulsar are magnified, causing a much greater tug than usual. Furthermore, the mutual interactions of the planets must influence the shapes and periods of each of their orbits in a predictable manner. If these effects could be measured, Rasio and his colleagues argued, then there would be no doubt that there are planets around the pulsar.

Wolszczan eagerly rose to the challenge and began to look for the predicted orbital changes due to the attractions of the planets for each other. For the next two years,

he collected additional data detailing arrival times of signals from the pulsar. All the while, he fed this information
into a computer, testing it against models of interacting
bodies. Armed with these new measurements, he convinced himself and his colleagues that the planets must be
real; their effects on the pulsars' broadcasts could not be
duplicated through any other realistic astronomical phenomena.

In April 1994, Wolszczan published a report in *Science*
announcing his confirmation of planets orbiting PSR
1257+12. He presented an analysis verifying conclusively
the existence of the two planets that he had found.
In addition, he presented strong evidence for a third,
smaller planet orbiting the pulsar in a tighter circle than
the other two.

In a decidedly conservative fashion, the planets were
named A, B, and C. (Wolszczan has stated that he has not
thought of more descriptive names for the new-found
worlds.) Planet A, the innermost of the lot, has a mass
similar to that of the Moon and orbits the pulsar once
every 25 days. The middle world, B, weighing in at 3.4
times the mass of Earth, has an orbital period of 66.5
days. Finally, the outermost, C, possessing 2.8 times the
Earth's mass, has a period of 98.2 days.

The new planets' orbits are paradigms of regularity.
Compared to Earth's elliptical (oval-shaped) excursion
around the Sun, they travel in near circles around their
dead central star. This monotony is only slightly affected
by the mutual tugs of the planets each time they near each
other.

The composition of Wolszczan's worlds cannot truly
be known until they are directly observed. Because of
their diminutive size, such prospects are most unlikely in
the foreseeable future. Nevertheless, Wolszczan has suggested that they are made of highly evolved stellar
material—complex elements built up over billions of

years of fusion. This matter, probably iron, was released from the core of the companion star destroyed by the pulsar. Thus Wolszczan suggests that the worlds he discovered are likely made of solid iron.

Most scientists believe that conditions on these pulsar planets would prove highly inhospitable for life. It is difficult to imagine living organisms being able to survive on what amounts to iron rocks continuously bathed in lethal radiation from an extinct star. Therefore, the finding of these novel worlds hardly represents new hope for the detection of extraterrestrial existence in space. Nevertheless, Wolszczan's work has paved the way for planetary discoveries of a far more promising nature.

As a pioneer in the radiotelescopic search for new planetary objects, Wolszczan has received considerable praise for his achievements. In December 1994 he was honored as a Grand Award winner by the editors of *Popular Science*. He continues to hold public lectures at Penn State, on radio astronomy, new planets, and prospects for extraterrestrial life.

Wolszczan continues on his quest for atypical pulsars and unusual phenomena surrounding them. Weaving his way between events and commitments in Puerto Rico, Pennsylvania, and Poland, he keeps himself quite busy with all of his interests and pursuits. A recent account of his research plans sounds most ambitious. As Wolszczan relates:

"There is always a lot to do. Perhaps the most exciting thing in progress is the search for superfast rotating pulsars (below 1 millisecond). This search may produce fascinating answers concerning superdense matter and quark stars. Another project which takes up a lot of my time has to do with the long-term studies of relativistic binary pulsars and tests of relativity theories using these objects as probes of gravity in strong field conditions."[6]

As if neutron stars weren't odd enough, it seems

that Wolszczan has embarked on a mission to find objects that are even more bizarre: pulsating stars made out of pure quark matter. Quarks (along with several other categories of particles) are the smallest known constituents of matter. They are ordinarily found within the cores of protons, neutrons, and similar subatomic objects. For example, each neutron is composed of three quarks. A quark star would be a tightly packed conglomerate of trillions and trillions of such minute particles. Because there would be no space between the quarks, it would be much denser than a neutron star. It would spin faster and have a stronger gravitational field within a more compact region. As it revolved it would emit pulsating beams of energy in intervals of less than 1 millisecond. Due to its extraordinarily high rate of rotation, it would probably be called a microsecond pulsar.

Could such a superfast pulsar be found in nature? Theory isn't certain. However, one might wager with great odds that if someone were to detect such a body— and maybe even find a planetary system around it—the discoverer would be a Polish Penn State professor working in Puerto Rico.

Chapter 5
BY JUPITER!
GIANT WORLDS
AROUND
ALIEN STARS

For high above your head your suns,
full and fulgurating, turn.
And yet, already in you is begun
something which longer than the suns shall burn.

Rainer Maria Rilke, *The Buddha in the Glory*

Planet-Hunting Season

Sometimes scientific discoveries trickle down slowly, like the unhurried dripping of a leaky faucet. Other times they gush in swiftly, like a sudden cloudburst on a hot summer night. In planetary astronomy, the year 1995 raged like a torrential thunderstorm.

At the beginning of that eventful year, there was absolutely no evidence that extrasolar planets existed around ordinary stars. Sure, scientists hoped that Wolszczan's work with pulsars could be extended to living suns. But, due to the troubled history of planet searching, most remained cautious and refused to predict when such worlds would be found.

By the end of 1995, however, solid proof had been found that at least four stars in our galaxy are orbited by planetary objects. Suddenly a whole new vista of a planet-speckled cosmos manifested itself to astronomers. The Sun, and its nine neighboring orbs, could no longer be considered a solitary system. From that point on, we could no longer suppose (if any lingering geocentrism existed in our hearts) that our corner of the universe was unique.

Most scientific revolutions are spurred on by a group of extraordinary leaders. As the first to find and confirm the existence of extrasolar planets, Wolszczan certainly proved himself a giant in his field. But there are those who deserve credit for other important achievements. For finding new planetary systems around living alien suns, considerable acclaim must go to the team of Geoffrey Marcy and Paul Butler.

Marcy and Butler have proven themselves to be the Holmes and Watson of planet-sleuths. With their keen eyes and sharp analytical skills, they have followed the trail of many dozens of potential planetary systems and have tracked down and/or confirmed six extrasolar worlds to date. Of these, five of the planets they discovered themselves, and one was a verification of another team's work. At this rate it will be a long time before their planet-hunting record is surpassed.

Geoffrey Marcy was born in Detroit in 1954 and grew up in the San Fernando Valley near Los Angeles. Like Wolszczan, he became interested in astronomy at a young age. When he was 14, his parents bought him a 4-inch reflecting telescope. He became fascinated by the nighttime spectacle that he could scrutinize with his instrument. Even so, he didn't anticipate his own success in the field of astronomy. As he relates:

"When growing up, I never imagined I'd find planets. Never!"[1]

Several years later, he enrolled at the University of California at Los Angeles, where he graduated with a double major in physics and astronomy. He went on to obtain a Ph.D. in Astronomy and Astrophysics from the University of California at Santa Cruz. Upon completing his education in 1982, he received a Carnegie Fellowship at the Mount Wilson and Las Campanas Observatories.

Returning to the San Francisco Bay area, Marcy was appointed to an astronomy position at San Francisco State

Figure 24. Geoffrey Marcy (b. 1954), co-discoverer of several extrasolar planets. (Courtesy of San Francisco State University)

Figure 25. Paul Butler, co-discoverer of several extrasolar planets. (Courtesy of San Francisco State University)

University, where he eventually rose to the status of full professor. There in 1987, along with postdoctoral researcher Paul Butler, he began to search in earnest for worlds outside of our own Solar System.

Marcy's quest was motivated by a long-standing interest in possible extraterrestrial habitats. As he describes this aspiration:

"I was inspired to search for planets due to a deep yearning to know if our Solar System is unusual or even unique. I also wonder whether our species, *H. sapiens*, is unusual in its fantastic intelligence, dexterity, and proclivity toward speech."[2]

Figure 26. Lick Observatory from the east. (Courtesy of UCO/Lick Observatory Photo/Image)

Researching with the 10-foot diameter Shane reflector telescope at the Lick Observatory on Mount Hamilton, California, Marcy and Butler started off by monitoring 60 of the brightest Sun-like stars visible from the northern hemisphere. Later they added 60 more targets to their list. By studying these stars for a number of years, they meticulously sought signs of Jupiter-size objects nearby. Their eventual success was clearly due to the painstaking nature of their efforts, as well as the potency of their search routine.

Doppler Detection

Marcy and Butler in their search used the spectroscopic method, which was pioneered in the 1980s by

Figure 27. The Shane telescope at Lick Observatory. (Courtesy of UCO/Lick Observatory Photo/Image)

Bruce Campbell of the University of Victoria, along with Gordon Walker and Stephenson Yang of the University of British Columbia. Using special detection equipment, attached to the 142-inch Canada-France-Hawaii telescope on Mauna Kea, Hawaii, the Canadian team conducted a

12 year hunt for extrasolar planets that ended in 1995. Although the Canadians' quest ultimately proved unsuccessful, their powerful approach strongly influenced the work of other investigative teams.

The spectroscopic surveys of these groups rely on the fact that the observed spectrum of light produced by a star (or other luminous celestial body) is affected by that star's motion through space. This phenomenon is an example of the Doppler effect: a shifting of a light source's apparent wavelength if it is moving with respect to an observer. If the source is traveling away from the viewer, then its light is shifted in wavelength toward the red end of the spectrum—that is, toward higher wavelengths. Green light transforms into yellow, yellow light into orange, and orange into red. On the other hand, if a source is coming closer to an onlooker, then its light is shifted toward lower wavelengths such as blue and violet. The same principle, as applied to sound, is the reason why the squeal of fire engines tends to rise in pitch as they approach and lower in frequency as they retreat.

One might readily imagine how the Doppler effect might be used by astronomers as a speed-trap to capture useful information about the velocities of stars. Like the devices police officers use to net speeders, Doppler-based detectors can be employed by researchers to see how fast particular stars are moving. To find out the speed of a star, they measure the observed shift in wavelength of its emitted radiation.

Unfortunately, a star's motion cannot be known completely by use of the Doppler technique. Only part of its movement, its radial velocity relative to Earth, can be discerned with this method. The radial velocity of a star is its speed along the line of sight from Earth. In other words, this measure signifies how fast a star is moving toward or away from us. Other aspects of a star's motion, for example its tendency to move up or down relative to

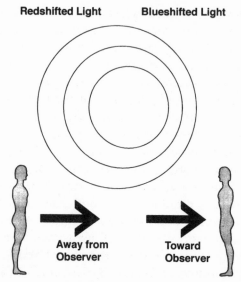

Redshifted Light   **Blueshifted Light**

Away from
Observer

Toward
Observer

Figure 28. In the Doppler effect, light from a source that is moving away from an observer appears shifted toward the red end of the spectrum (toward longer wavelengths). On the other hand, light from a source·that is coming toward an observer appears shifted toward the blue end of the spectrum (toward shorter wavelengths).

the plane of our galaxy, or its movement across the line of sight from Earth, cannot directly be found using this procedure.

Still, the value of the radial velocity of a star, if known with enough precision, provides a vital clue as to whether or not a planetary system is present around that body. As we have seen in our examination of the wobble method pioneered by van de Kamp, and later developed by Gatewood, planets exert a small, but measurable (at least theoretically), gravitational tug on the stars they orbit. In the case of the wobble method, astronomers try to discern this pull by recording the changes in stellar positions. However, the gravitational influence of planets also affects the speed of stars. By measuring changes in stars' radial velocities, astronomers can, in principle, find out if these shining sources are ringed by unseen objects.

Thus, the main difference between the wobble technique and the spectroscopic method lies in whether stellar positions or speeds are used as gauges of the possible gravitational effects of planets. With the former approach, astronomers must pinpoint exactly where a star *is*. With the latter they must track exactly how fast it *goes*.

One might, by analogy, imagine applying these two methods to a visible female dancer (a star) engaged in a waltz with an invisible man (a planet). Suppose that one were watching such a dancer whirl around and were interested in the strength and dexterity of the unseen dance partner who is leading her. That is, one wants measure his pull on her by means of observing her movements. Is her invisible partner a slight, frail man, for instance, or a heavy, brawny man?

The wobble approach to this dilemma would involve tracking the dancer's footsteps. By observing where she steps for each beat, the twirls and tugs of her hidden partner might be guessed and estimates of his strength and dexterity be made. In contrast, the velocity method would involve seeing how fast she moves, back and forth, as she swirls and pirouettes across the floor. Then one might guess how powerful and mobile he would have to be to swing her around at such speeds. To estimate this, one would have to make certain assumptions about how much of her dancing is due to her own volition, and how much is the effect of being led by her shadowy companion. If that were known, a pretty good picture could be constructed of the partner's strength and agility. In analogous fashion, both the wobble and spectroscopic techniques can be applied to studying a possible unseen, but influential, mate of a star.

Each method requires great amounts of precision so the decision as to which to use has traditionally been a difficult one for researchers—a matter of personal preference. Recent successes using Doppler-tracking (i.e., the radial velocity method), however, have led many into the spectroscopic camp. Speed-trapping, rather than wobble-watching, has snared the recent bonanza of planetary finds. Thus the balance seems to be tipped at present in the direction of measuring radial velocity, rather than position, to find new objects.

Part of this movement toward the use of spectroscopy can be attributed to improved methods. Over the past decades velocity techniques have become significantly more accurate. Before Campbell, Walker, and Yang began their survey, astronomers would consider themselves lucky if they could measure radial velocities within a precision of several hundred feet per second. But this level of certainty clearly wasn't enough to find planets. Planetary tugs on stars are too subtle to be recorded with such imprecise techniques. For instance, Jupiter's gravitational pull on the Sun causes it to move back and forth with a radial velocity of merely 40 feet per second. This introduces a Doppler shift in the wavelength of light from the Sun of less than three parts per 10 million. Thus early spectroscopic instruments possessed nowhere near the power to see a large planet of Jovian proportions, let alone one the size of Earth. Using these devices to search for new worlds was like trying to study microbes with a magnifying glass found in a cereal box.

With their updated research program, the Canadian team changed all this. They designed equipment and perfected techniques to bring their accuracy within 45 feet per second. For the first time in scientific history, their methods offered the possibility of placing large planetary bodies within range of detection.

The means they used to measure the Doppler shifts of starlight depended on comparing stellar spectra to the characteristic patterns of a known source. The chemical yardstick they used was the wavelength distribution of hydrogen fluoride gas. All light from the telescope, before it was analyzed, passed through a cell of this substance. Then the wave patterns of the starlight and the gas were matched up against each other.

The reason Campbell, Walker, and Yang performed such a comparison was that they expected any instrumental errors affecting the starlight's spectral lines would similarly affect the appearance of the hydrogen fluoride's spectrum. Any fluctuations rendered by environmental distortions of the stellar light would make their presence known in the wavelength patterns seen in the gas. Thus, the group could readily detect and then cancel out such errors. This correction mechanism—akin to horizontal etchings on the side of a glass of water that would indicate when it was tilted (by the surface of the water no longer lining up with the etchings)—greatly increased the precision of their work above that of previous endeavors.

Apparently even the impressively high precision of the Canadians' apparatus wasn't sharp enough for them to find planets. They spent over a decade of careful work monitoring 18 stars similar to the Sun. Of these 18, nine showed significant motion. There were some indications that an unseen object was affecting the velocity of the star Epsilon Eridani. However, none of the patterns of behavior indicated without question that planets were present. Rather, these spectral shifts could each be equally well attributed to small random stellar perturbations. Moreover, independent groups did not confirm the existence of any of these stellar movements. Therefore, at the close of their project, Campbell and his colleagues could not claim success.

When Marcy and Butler decided to improve on the Canadians' approach, they elected to replace the use of a hydrogen fluoride compartment with the employment of a cell filled with molecular iodine. Other teams engaged in spectroscopic quests, such as the group lead by William Cochran at the University of Texas McDonald Observatory, similarly chose to utilize iodine as a calibrator (source of comparison). This replacement was partly due to the volatile nature of hydrogen fluoride. Hydrogen fluoride is an extremely reactive substance known to cause chemical burns if spilled. For this reason, iodine was welcomed by astronomers as a much less hazardous material to use.

More important, iodine offered researchers an excellent reference spectrum. Its spectral lines are very sharp in the wavelength region in which much stellar light can be found. Therefore, by matching up its spectrum against that of incoming starlight, it provides a superb gauge of possible instrumental distortions.

By using the iodine cell as a calibration tool, as well as using an upgraded ultrasensitive electronic camera and superior computer modeling programs, Marcy and Butler managed to obtain consistent radial velocity errors of less than 10 feet per second. Thus, their equipment could record stellar data in a manner more than four times as precise as that of previous groups. This placed their group in an enviable position to scout out any new planets to be found.

Bellerophon, the Horseman of Pegasus

In Greek mythology, Pegasus was the beautiful winged horse who, with a kick of his hoof, caused the stream Hippocrene to flow from Mount Helicon. Later he thundered through the sky as Zeus' flying steed. According to the ancient Greeks, that is why we can now see Pegasus among the stars. Those with a keen imagination can trace out the shape of the horse in the heavens, as outlined in the constellation known as the Great Square of Pegasus.

Recent events have brought the Pegasus formation added significance. The Great Square is the site of the first discovered extrasolar planet around a living star. What a grand part of the heavens for the finding of a new world—in the kingdom of the flying horse.

It is hard to think of a more fitting image to represent astronomy's quest for planets than that of a thundering steed sprinting toward new horizons. The hunt for planets has truly been a horse race. Many teams have stampeded ahead in their quest for the grand prize of capturing the first extrasolar world, only later to be set back by unforeseen problems. A winged horse, able to leap these hurdles, would have come in handy. But sometimes inextricable circumstances, such as equipment limitations, force astronomical groups to plod along in the mire, barely able to trot, let alone spring forward.

Naturally, in any horse race, there are those who win, place, and show. A good gambler divides his wager among all of these categories, divvying it between them as the odds dictate. One never knows what powerful steed might come in second or third one day, while another equally worthy contender noses ahead at the last minute. Accordingly, sometimes it pays to be conservative and hedge one's bets.

Some might have bet that Marcy and Butler, with their exceptionally precise equipment and strong power of will, would have been the ones to find the first extrasolar planet around a living star. Not so. They were the ones to find the second world, the third, and a number of others. But, much to their disappointment, they were scooped when it came to the first.

It was a Swiss team who, in 1995, scooped up evidence of the first known sun-circling extrasolar world and scattered this information across global headlines. In a conference that year in Florence, Italy, Michel Mayor of the Geneva Observatory and his graduate student Didier Queloz announced their discovery of a planet around the star 51 Pegasi. Strongly resembling the Sun, 51 Pegasi is situated in the heavens near one of the edges of the Great Square and lies only 40 light-years away from Earth.

Ironically, Mayor and Queloz weren't mainly looking for extrasolar worlds. They were primarily engaged in an effort to find brown dwarfs—low-mass stars that had failed to ignite. Their equipment, possessing a sensitivity of 40 feet per second, was tuned to a much less precise degree than that of Marcy and Butler, calibrated within 10 feet per second. Yet the Geneva researchers managed to scope out a prime candidate that had somehow evaded the San Francisco group.

Marcy and Butler say they certainly would have included 51 Pegasi in their search, if it weren't for the fact that it was miscatalogued. Because the Yale *Bright Star Catalogue* lists it as a subgiant star, rather than as a Sun-like star, they thought it would be of little interest for planet-hunting. Subgiants tend to be unstable and therefore rather poor candidates for harboring living worlds. Like a merry-go-round with a decaying axis, they do not provide steady central supports for revolving objects. The team decided that looking for a planet around a star so different from ours would be a waste of time when there are so many bodies in our Galaxy more akin to the Sun. Later on, when the San Francisco group realized that 51 Pegasi is actually virtually identical to the Sun, they were sorry they had excluded it.

Mayor and Queloz, in contrast, did not have grandiose hopes for finding close matches to our own Solar System. Rather, they were tracking a broad range of single stars and looking to see if any of these had an invisible companion of either the brown dwarf or planetary variety. Therefore, 51 Pegasi, known to be relatively close to

Earth and single, was included in their search, along with 141 other selected objects of interest.

In April 1994, they began tracking the radial velocity of 51 Pegasi and the other target stars. With a state-of-the-art spectrograph attached to the 6-foot-diameter telescope at Haute-Provence Observatory in France, they ardently plotted how each of these bodies waltzed back and forth along the instrument's line of sight. By plotting these stellar promenades they hoped to glean evidence of invisible dance partners.

By October 1995 they had gathered enough data to announce success. At the Florence workshop on cool stars, they reported strong evidence that a planet half the mass of Jupiter is engaged in a circular orbit around 51 Pegasi. Triumphantly, they heralded the first world found around an alien star similar to the Sun. Following astrophysical convention, they named it 51 Pegasi B.

The data Mayor and Queloz presented, gathered from over a year's worth of observations, showed that 51 Pegasi seemed to be oscillating back and forth with respect to Earth with a maximum speed of about 200 feet per second each way. Technically, this was indicated by the semiamplitude (height, using zero as a basis) of the radial velocity versus time curve that they plotted for the star. Because of the dimensions of this curve, it was clear that a massive object of planetary proportions was gravitationally affecting the star.

The same graph showed that the planet orbits 51 Pegasi at an incredibly fast rate of approximately once every 4¼ days. This rapid revolution implies that the distance between the star and planet is only about 10 million miles—eight times closer than Mercury is to the Sun. This is a fantastically close distance for such a large world, much nearer than once believed possible.

Right before the conference, Mayor had contacted several theorists to check if the laws of astrophysics permitted such a body to exist so close to a star in a stable orbit. He wondered if a planet in that position would continue to revolve indefinitely, or would soon spiral into the central fire of its mother star. None of the astrophysicists he asked had been sure. Then a former student of Mayor's spoke with theorist Adam Burrows of the University of Arizona about the problem. Burrows ran some computer simulations, showing that such an orbit would indeed be stable. After Mayor looked over Burrows's results, he felt much more comfortable presenting his discovery to the conference.

The analysis performed by Burrows showed that an object could, in theory, exist where 51 Pegasi B was found. However, it didn't explain exactly how such a celestial titan came to be there. How could a world half the mass of Jupiter be formed so close to its sun?

Since it was hard for planetary astronomers to fathom such a situation, many were incredulous at first about Mayor and Queloz's discovery. No theory of planetary formation could account for such proximity. As in the Solar System, large extrasolar worlds were thought to be formed of ice and rock, far away from the fires of their suns. Only smaller planets, fashioned from more durable materials such as iron, were thought to reside within the inner zones of planetary systems. For this reason, many astronomers, including Marcy and Butler, expected at first that the planet sighting would turn out to be a false alarm.

Marcy and Butler decided to test the matter themselves. They took advantage of four nights of reserved telescope time to conduct a full investigation of whether or not there is a planet around 51 Pegasi. Four nights may not seem very long, but it constituted nearly a full year for the planet (assuming that it existed). With their high-precision equipment, the San Francisco team plotted the radial velocity of the star over this period. They were

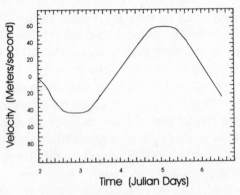

Figure 29. To confirm Mayor and Queloz's discovery of a planet around 51 Pegasi, Geoffrey Marcy plotted the star's radial velocity as a function of time. Sure enough, he noticed a distinct sinusoidal variation in the velocity versus time curve, indicating the tugging effect of an unseen planet.

elated to discover that their data formed a perfect sine wave, confirming the presence of a planet in circular orbit around 51 Pegasi. Mayor and Queloz were right; 51 Pegasi B is a genuine extrasolar world.

The fact that Marcy and Butler determined the orbit of 51 Pegasi B to be a geometric circle, rather than an ellipse, lent further weight to the proof. According to astrophysical theory any planet so close to its own sun must follow such a path. The reason for this is that stellar tides tend to have a smoothing-out effect on the motions of orbiting bodies. For an object revolving around 51 Pegasi at the proximate distance of only 10 million miles away, these tidal forces would be immense—enough to round out any deviation from a perfect circle. Thus, the ring-shaped path observed for the Pegasus planet is exactly the type of motion that astrophysicists expected.

In addition to confirming the existence of a planet and predicting its trajectory, Marcy and Butler verified that its mass is likely half that of Jupiter. However their results carried with them a margin of error; the planet could be lighter or heavier than estimated.

This uncertainty about the mass of the newly discovered world stems from not knowing the tilt of its orbital plane relative to its line of sight from Earth, a physical parameter called the inclination. Astronomers' estimated value of the planet's mass assumes that they have been viewing the orbital plane at an angle of 20 degrees from edge-on. This small angle represents their best guess as to the orientation of the 51 Pegasi system relative to us. But it is only an supposition, and could be wrong.

If, instead, they have been observing the planet in a direction perpendicular to its orbit, then they would have been greatly underestimating its mass. In that case, the measured radial velocity of 51 Pegasi would have represented but a minute part of its total motion due to the planet. A large element of the star's movement, that perpendicular to the line of sight, would have remained undetected. Thus, the mass estimate gleaned would have comprised only a fraction of a much greater true value. Consequently, in the unlikely event that astronomers have been viewing the system at a large angle from its orbital plane, the planet around 51 Pegasi could conceivably be dozens of times the mass of Jupiter.

By the time Marcy and Butler completed their analysis, the bulk of the astronomical community had become convinced of the reality of 51 Pegasi B and of the essential nature of its properties. Along with Mayor and Queloz, Marcy and Butler were heralded internationally for work well done. Receiving hundreds of letters and electronic mail messages per week from journalists, excited school children, UFO enthusiasts, and other followers of science, both groups quickly became deluged with publicity. Clearly, their research tapped into a wellspring of public

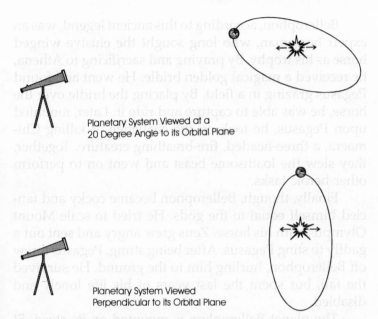

Planetary System Viewed at a
20 Degree Angle to its Orbital Plane

Planetary System Viewed
Perpendicular to its Orbital Plane

Figure 30. When astronomers view a candidate planetary system, they cannot directly determine the tilt angle of the system with respect to the line of sight. Estimates for the mass of 51 Pegasi B are based on a tilt angle of 20 degrees. However, if the tilt angle is much larger, mass estimates of the planet would have to be similarly increased. This is because the greater gravitational tug of a more massive planet would be required to produce the same observed radial velocity.

interest in extrasolar planets and extraterrestrial life. As Geoffrey Marcy relates:

"The media attention has been overwhelming. Paul and I can barely find time to work on our research, with all the TV, radio, and newspapers interviewing us."[3]

As a further contribution to their planet-hunting endeavor, Marcy and Butler suggested an intriguing name for the new-found world. They dubbed it "Bellerophon," after the Greek mythological hero who snared Pegasus and rode it to glory.

Bellerophon, according to this ancient legend, was an expert horseman, who long sought the elusive winged horse as his trophy. By praying and sacrificing to Athena, he received a magical golden bridle. He went and found Pegasus grazing in a field. By placing the bridle over the horse, he was able to capture and ride it. Later, mounted upon Pegasus, he took on the challenge of killing Chimaera, a three-headed, fire-breathing creature. Together, they slew the loathsome beast and went on to perform other heroic tasks.

Finally, though, Bellerophon became cocky and fancied himself equal to the gods. He tried to scale Mount Olympus with his horse. Zeus grew angry and sent out a gadfly to sting Pegasus. After being stung, Pegasus threw off Bellerophon, hurling him to the ground. He survived the fall, but spent the last years of his life lonely and disabled.

The planet Bellerophon is mounted on its steed, 51 Pegasi, rather closely. The gravitational harness connecting them is only about 10 million miles long—rather tight reigns for a planetary system. What, then, does this proximity indicate about the type of world Bellerophon represents?

In a word, the planet around 51 Pegasi is hot. Red hot, to be exact. Its nearness to the scorching heat of its star causes its surface temperature to remain on average around 1000 degrees Celsius (about 1800 degrees Fahrenheit). Without a doubt, no liquid water, believed to be essential for living organisms, could remain on the planet's surface. Thus, it would be folly to expect to find flourishing life on that world. No alien civilization, or even microbial colony, could survive such hellish conditions.

The hero Bellerophon should have thanked Zeus for his fate. It could have been far worse. He could have been condemned to live out his days on his namesake planet.

Temperate New Territories

Not to be outdone by the Swiss, Marcy and Butler spent the months after the discovery of 51 Pegasi B fervently looking through their own collected data for novel worlds around additional stars. They hoped that their extensive sky-scan, using their high-precision equipment and computational facilities at Lick Observatory, would snare them some planets they could claim as their own finds. Confirmation is important, but the thrill of the hunt lies in the discovery.

Aside from the challenge of finding new worlds with his own instrumentation and computers, Marcy wanted to prove that the sighting of 51 Pegasi B was not just a fluke. If planets are common throughout the cosmos, then others should be spotted as well. The discovery of the Pegasus planet could not just be a one in a million event. Surely, he thought, there were more yet to be found.

Moreover, Marcy had a deep desire to find worlds in space that could possibly support living beings. This was one of the main reasons he began his search in the first place. He wanted to explore new realms that bore similarity to Earth. It was obvious to him that 51 Pegasus B was not such a place. Therefore, he sought more promising candidates for habitability.

In January 1996, at the Texas meeting of the American Astronomical Society, Marcy and Butler made global headlines with their announcement that they had found the first worlds located in the temperate zones of their systems. They reported the discovery of two new planets: one around the star 70 Virginis, in the constellation Virgo, and the other around 47 Ursae Majoris, in the constellation that contains the Big Dipper. Unlike 51 Pegasi B, these worlds occupy comfortable distances from their suns— far enough away to support liquid water.

The planet around 70 Virginis, for instance, orbits at

half the distance of Earth from the Sun. Its surface temperature is estimated to be about 80 degrees Celsius (176 F). Thus, it is hot, but not boiling hot; a sauna, not a frying pan.

The Big Dipper object revolves around its star at a somewhat farther distance than Mars from the Sun. Its orbital radius is twice that of Earth's. Therefore, it is probably much colder than Earth and almost certainly more frigid than the Virgo planet. Yet, it still occupies a relatively temperate position in its system.

The fact that water might flow on these worlds lends itself to suggestions of habitability. Marcy fostered these speculations by picturing the planets as spawning grounds for complex organic molecules, amino acids, and perhaps even primitive organisms. Some in the media took this several steps further by painting scenarios of advanced alien creatures roaming the exotic expanses of the newly detected bodies. These far-fetched pictures were reminiscent, to some extent, of the green men on Mars theories during the era of Schiaparelli and Lowell and similar ponderings about Barnard's star during the time of van de Kamp.

However, these fanciful speculations about 70 Virginis and 47 Ursae Majoris soon became replaced with more sober assessments. Just because celestial objects possess the correct temperature to support liquid water doesn't mean they are covered with oceans, lakes, and rivers. They could just as well have dry surfaces like Mars or Venus. (True, there is evidence that water once flowed on Mars, but that needn't be the case for the new planets.) The additional tests necessary to establish whether or not these planets possess water lie far in the future.

Furthermore, each of the worlds found by Marcy and Butler is quite massive compared to Earth. The planet around 47 Ursae Majoris has a mass at least three times that of Jupiter. Therefore, its surface gravity is extremely strong. Organisms living there would have to be well adjusted to weighing thousands of times more than they would on Earth. Perhaps no living creature could survive such pressing force—equivalent to swallowing a hundred lead sandwiches, smothered with cobalt dressing, and generously topped with uranium chips.

The object near 70 Virginis is even heavier. By conservative estimates, it weighs at least nine times as much as Jupiter. Because of its great bulk, many astronomers think it is a brown dwarf rather than an actual planet. Brown dwarfs, being failed stars, probably don't support life.

The brown dwarf hypothesis is bolstered further by the planet's stretched-out orbit. Instead of orbiting its star in a circular or near-circular path (as planets do in our own Solar System), it revolves in an elongated oval. Recall that astronomers refer to a deviation from perfect circularity by the term eccentricity. By this measure, the planet around 70 Virginis follows a highly eccentric trajectory.

Regardless of their habitability, the discovery of the two new objects by the San Francisco team represented a great step forward in the chronicle of planet-hunting. Marcy and Butler showed that worlds around Sun-like stars are emphatically not rare. Rather, they seem to be scattered among the stars like coconut trees on South Seas islands. This apparent abundance is good news for those seeking intelligent life in the cosmos. It means that we may not have to look far to find cousins to the Solar System.

Novel and Eccentric Worlds

Marcy and Butler wasted no time in delivering news of additional planetary discoveries. In April 1996, they reported the observation of a world orbiting Rho Cancri, a star in the constellation of Cancer. This planetary object was found to have an extremely close orbit—about ⅒ the radius of Earth's trajectory around the Sun. In June 1996, they announced the finding of a second planet around the same star. These sightings represented the first planetary system—of two objects or more—found around a star other than the Sun.

Also in June, at the Madison meeting of the American Astronomical Society, the San Francisco team and the Geneva group led by Mayor reported independent discoveries of a planet near the star Tau Bootes. It was found to have a mass of at least four times that of Jupiter, and have an orbital radius of about the same distance of that of 51 Pegasus B. Shortly thereafter, Marcy and Butler, along with several other astronomers, presented a paper documenting evidence of yet another planet, near Upsilon Andromedae, with mass and orbit similar to the Pegasus object.

That October, two independent groups—Marcy and Butler and the University of Texas (McDonald Observatory) team of William Cochran and Art Hatzes—announced they had each found evidence of a planet around the star 16 Cygni B. Then, in April 1997, another team found a Jupiter-like body orbiting Rho Coronae Borealis, a star in the Northern Crown constellation. This brought to nine the tally for planets detected by means of their stars' radial velocities, and to over a dozen the possible number found by all methods, including the three pulsar planets discovered by Wolszczan, another pulsar planet, around PSR B1620-20, announced by Z. Arzoumanian and his colleagues in May 1996, and two objects near the star Lalande 21185, reported in June 1996 by Gatewood. Not all of these sightings have been confirmed.

For the first time in history, the quantity of known planets outside of the Solar System is on par with (or perhaps greater than) the number of those inside its bounds. Giardano Bruno's prediction that other suns have other worlds has finally been realized. No longer cloistered in our own tiny part of space, we have broken free of our system's cocoon and enacted the first efforts to reach out and observe alien locales. We are still alone, as far as we now know, but we can no longer claim ignorance.

In comparing the newly found worlds to our own system, many questions remain. Why are the orbits of some of these planets so close to their stars? Why do others revolve in such eccentric, oblong fashions, rather than along paths closer to circles? In short, why do the recently discovered systems seem so different from our own?

The planets around 51 Pegasi, Rho Cancri, Rho Coronae Borealis, Tau Bootes, and Upsilon Andromedae, for instance, orbit these stars at a stone's throw. They occupy places where, in our own Solar System, no physical objects exist. Their existence stands in sharp contradiction to known planetary theory, which presumes that any such objects so close to their stars would get swallowed up. Moreover, as we've discussed, Jupiter-size planets were thought to be created out of the ices common to the outer regions of planetary systems, rather than out of the rocks and dust prevalent in the inner zones. In each of these examples, astronomers have found mammoth worlds virtually as close to their central stars as they can stably reside.

In two of the other discoveries, the detection of objects around 70 Virginis and 16 Cygni B, researchers have tracked planetary orbits that are extraordinarily eccentric. Eccentricity, a measure of how stretched out an orbit is, ranges in value from 0 to 1. The lowest value applies to a purely circular path; the highest to an open orbit, one in which the revolving object flies off into space. Planets in the Solar System have very small eccentricities, ranging from approximately 0.0, in the case of Earth and Venus, to more than 0.2, in the case of Pluto. Most have eccentricities of less than 0.1. In contrast to these low figures, the eccentricity of the world around 70 Virginis was found to be 0.4, and that of the 16 Cygni B planet a whopping 0.6. Rather than revolving regularly around their suns like the grooves of a constantly spinning record, they leap inward and outward like skipping phonograph needles. One might expect such behavior from wandering comets, but hardly from stable, full-fledged planetary bodies.

Life on such eccentric worlds would be difficult at best and probably impossible. Because of their erratic paths, their surface temperatures would oscillate wildly from season to season. During their close passes around their suns, exterior conditions would be boiling hot. Later, when the planet flees to the outer zone of the system, everything on the surface would freeze up. Still, unlike the perpetually scorching environments on star-hugging worlds such as 51 Pegasi B, there would be at least some intervals of temperate weather, between the two extreme periods. Perhaps organisms might evolve on an eccentric planet equipped to deal with these temperature variations, thriving during the balmy periods and battening down when frost or fire arrives. Or maybe life there would never have a chance to form.

There are several possible explanations for the errant motions of the newly found celestial objects. The simplest, but most discouraging, possibility is that at least some of these bodies are not planets at all, but rather constitute brown dwarfs. Because brown dwarfs form independently, rather than as part of planetary systems, there is no constraint on how close they are to other stars or how they orbit.

As we've discussed, planetary systems are believed to be created, along with their stars, from rotating protoplanetary disks. During such formation processes, the centers of the disks coalesce into stars, the dusty inner shells into the terrestrial planets and the icy outer reaches into the Jovian worlds. The distinctive nature of this evolution places constraints on planetary sizes and behaviors.

A brown dwarf, on the other hand, might conceivably form in space and then end up, by coincidence, close to a Sun-like star. Because it did not form along with the star, it might potentially be engaged in any of a wide range of orbits, spanning a broad spectrum of possible distances and eccentricities. Thus a massive object, such as that circling 70 Virginis, might fit the profile of a failed star, rather than of a true planet. (And, as discussed, its mass might also put it into that category.)

It would be disheartening indeed if it does turn out that many of the objects found by Mayor, Marcy, Cochran, and the others are brown dwarfs, rather than planets. Astronomers' estimates of the number of true planetary systems in space would then need to be reevaluated. They would need to work harder, with equipment even more precise than already used, in order to seek out genuine orbiting worlds.

However, aside from being brown dwarfs falsely mistaken for planets, there are other reasons why the new objects might behave so erratically. Perhaps they are genuine planets that have had their original orbits disrupted. Although they started out in their systems traveling along circular paths, relatively distant from their mother stars (i.e., the distance of Jupiter or Saturn from the Sun), cosmic catastrophes caused them to spiral inward. For example, maybe collisions with intervening objects—proximate stars, other planetary bodies, or large chunks of interstellar debris—served to reroute the planets' trajectories. In some cases, these impacts caused the worlds to assume orbits much closer to their suns, in others, to redirect themselves into highly eccentric paths. This would explain why the paths of some of the new-found worlds are so tight, and of others, so oblong.

Another possibility is that the fledgling planets were

drawn inward during their formation processes. Perhaps in some systems dust and rocks in the protoplanetary disks attracted, through their combined gravitational pull, Jupiter-size objects inward. If this was the case, the question remains, why did this process come to a halt? Why weren't the Jovians flung into their suns? Theorist Doug Lin, of the University of California at Santa Cruz, has been working on this difficult problem. So far, a satisfactory solution has yet to be developed.

Researcher William Cochran, co-discoverer of the 16 Cygni B planet, has advanced another option. He has argued that it is circular reasoning to assume that all planetary systems resemble ours. It is ridiculous, he believes, to surmise that all "normal" planets must orbit in near-circles, just because our own Solar System's worlds do.

Imagine if a hermit spent his entire life (to that point) on an tropical island in which all the birds had golden beaks and purple plumage. He might very well assume that these colors characterize every flying creature. He'd be most surprised, then, if he were to leave his sanctuary for the mainland and come face-to-face with red-breasted robins, white doves, and grey pigeons. He would realize that his earlier assumptions were misguided. With similar reasoning, Cochran asserts that astronomers would be wrong to suppose that all planetary systems are carbon copies of ours.

In any scientific endeavor, developing good statistics requires a large sample. So far, with only a dozen or so worlds found to date, we certainly should not generalize about the nature of all planetary systems. To establish the truth, many more examples of extrasolar objects are needed.

To that end, Marcy and Butler press on in their painstaking search for alien planets. As the Holmes and Watson (or perhaps Crick and Watson) of planetary discovery, they have decoded the murky messages hidden in the subtle movements of distant stars, revealing the unseen sources of these motions. Together, they hope to unravel many more of the galaxy's dark secrets. And given their fine track record, there's reason to believe they will.

Perhaps the secret to the Californians' success is their teamwork. They function well as a unit because, as Marcy relates:

"I enjoy working with Paul Butler because he is extremely clever, witty, and original. He compliments my style by working fast and getting things done while I tend to cogitate slowly."[4]

Judging by Marcy's description of their daily routine, such compatibility is needed for the grueling hours they put in together (of which Marcy's hard-working wife, chemist Susan Kegley of Berkeley, is quite supportive):

"At Lick Observatory, we work very hard—usually 17 hours per day. We start at about 2 PM to set up all the computers, spectrometers, telescopes, and to determine the target list of stars. Then we work all night until about 7 AM, when we finally go to sleep. It is wonderfully exciting to work all night, as the data pour in!"[5]

And it is the data collected by researchers such as Marcy, Butler, Mayor, Cochran, and the like upon which our hopes for interstellar exploration are built. Their nightly labors, mapping out remote territories upon which no human footsteps have trod, form the true cartography of the future.

Seeing the Invisible

Even the boldest endeavors of the Swiss, Californian, and Texan teams have well-defined limitations. Because of the built-in constraints of their instruments, these groups do not expect to detect Earth-size objects or to photograph directly planets of any proportions. Their Doppler-based equipment shares one specific purpose: to measure the wobbles of stars caused by nearby "jupiters." Imaging other "earths" lies beyond the scope of their missions.

Clearly, the search for life in space requires more. We must develop the means to record the faint impressions of terrestrial worlds. Because of their moderate gravitational fields and presumably rocky compositions, Earth-size planets represent the best candidates for habitability. And to find out if such bodies physically possess particular features suitable for life, we must be able to see them.

In a recent speech to the American Astronomical Society, Daniel Goldin, the head of NASA, indicated profound optimism about the possibility of being able to produce visual images of extrasolar worlds:

"In 25 years," he said, "we'll be able to image an Earth-sized planet and see ocean, clouds and mountain ranges ... That is a central goal for NASA—to find out whether Earth is unique."[6]

One hopes that Goldin's forecast proves correct. Yet many obstacles remain before astronomers might directly view planets. Naturally, the main stumbling block is the relative faintness of planetary objects compared to stars. Starlight shines with billions of times the intensity of planetary luminance. One might just as well hope to see a glowworm in an erupting volcano as to observe a world next to its sun.

The physical imperfections of current optical instruments make matters much worse. Tiny flaws in telescopic mirrors scramble the light from stars, forming haloes that can drown the feeble light of surrounding planets. Moreover, even if these defects are reduced, diffraction causes starlight to appear spread out into light and dark bands, smearing out the visual zone in which planets might be seen. All in all, it would be extremely hard to make out the image of a planet among all the scattered light.

Standard, ground-based telescopy must cope with yet another obstacle: the distorting effects of Earth's atmosphere. Heat pockets in the air affect the speed and direction of light arriving from space. Cooler parts of the atmosphere transmit light in a different manner than warmer regions. Consequently, all stars seem to twinkle when viewed from the ground.

Recently, Roger Angel, an astronomer at the University of Arizona, proposed using the techniques of adaptive optics to perform ground-based direct imaging of extrasolar planets. (Recall that these methods have been suggested for astrometry as well.) By using rubber mirrors (large reflectors built from many small, mobile reflective surfaces), he has shown how atmospheric blurring might be reduced significantly. In his scheme, special computer programs would measure atmospheric distortions, and then steer the tiny mirrors to correct for them. Each time atmospheric conditions were found to be altered, the rubber mirror would accordingly be bent to compensate.

Angel believes that his method, applied to telescopes with mirrors at least seven feet across, would be suitable for the direct imaging of Jupiter-size extrasolar worlds. Eventually, perhaps, even Earth-size planets might be found in this manner. But Angel feels that in the near future the chances of viewing terrestrial worlds around distant stars with ground-based equipment are remote. Even the best adaptive optic techniques currently available could not provide the sharp focus required to image such faint targets.

Naturally, one way of avoiding the conundrums of ground-based telescopy would be to take measurements from space. The Hubble space telescope was originally designed with the intent of astronomers being able to use it to image large extrasolar planets. Since observation time on the Hubble is so dear, it hasn't so far been employed toward that aim.

Looking with the Hubble for Earth-size extrasolar worlds would be futile. A much larger space instrument, with a far smoother mirror, would be required to perform this task. It is estimated that a telescope with a mirror about twice the diameter and ten times as even as the Hubble's would be able to image the closest terrestrial planets beyond the Solar System.

Such a mammoth space telescope would be extremely expensive. Considering today's tight governmental budgets, it is unlikely to be built for many years. Until then, we must patiently await the first images of alien worlds similar to our own.

Chapter 6
HIDDEN WORLDS: PLANETS AS DARK MATTER

Darkness settles on roofs and walls,
But the sea, the sea in the darkness calls;
The little waves, with their soft, white hands,
Efface the footprints in the sands.
And the tide rises, the tide falls.

HENRY WADSWORTH LONGFELLOW, *The Tide*
Rises, the Tide Falls

The Hidden Majority

Stars are not shy. They make their presence known for trillions of miles around them by the generous stream of radiation that they cast off each second into space. Starlight is so brilliant that, even with the naked eye, tens of thousands of shining orbs can be observed in the clear nighttime sky.

Planets, in contrast, are more demure. Like bashful children, they hide within the folds of their mother suns' glowing skirts. Stellar radiation cloaks them from the inquisitive glances of scientists in such an effective manner that, in the many centuries since the invention of the telescope, none beyond our Solar System have directly been seen. The dim visages of planetary bodies are simply overwhelmed by the dazzling output of luminous spheres.

There are myriads of shining objects in space. In the Milky Way alone there are approximately 100 billion stars. This vast collection includes Sun-like orbs beaming orange and yellow, bloated giants blushing blue and red, exploding supernovas, and many other varieties of light sources. Countless additional stellar bodies are grouped into billions and billions of galaxies. We observe these as spirals, ellipticals (ovals), and other brilliant shapes, speckled throughout the celestial dome. And the stars and galaxies hardly represent the only luminous entities in the sky. Astronomers have observed powerful beacons of intense light, called quasars, that outshine many galaxies. They have found large numbers of these scattered about the farthest reaches of the universe. All in all—when one tallies stars, galaxies, quasars, and other astronomical objects—it is absolutely staggering to contemplate the sheer quantity of glimmering sources in the cosmos.

Considering the vast number of bright objects in the universe, the scientific community has traditionally believed that the bulk of it is visible. Until recently, the preponderance of astronomers thought that with a powerful enough optical instrument they could, in theory, map out most of the universe. They reasonably assumed that most of the matter in space is luminous.

However, in the past few decades, overwhelming evidence has accumulated, indicating that the bulk of the material in the cosmos (at least 90 percent and perhaps over 99 percent of its mass) is invisible. That is, most of the universe is made of dark matter undetectable by ordinary telescope. No matter how carefully astronomers have searched for these hidden objects and substances, their sky surveys have come up short. There is much to the cosmos that cannot be seen, but can only be inferred.

Scientists believe that a certain percentage of this hidden matter resides in the form of unseen new bodies. Planets and faint stars—of the white, red, and brown dwarf varieties—form a significant part of the unseen material in space. Therefore the hunt for planets by astronomers is also a search for dark matter. Each new planetary object found provides insight into the nature of the invisible portion of the universe.

Our current map of the cosmos is like a great jigsaw puzzle. The more obvious pieces of the picture, the stars, galaxies, and other brilliant objects, have, for the most part, already been put into place. Discoveries of novel worlds help researchers to assemble the parts of the puzzle that are less apparent. Eventually, they hope with this and other information to fill in the obscure segments of the map, and complete their portrait of the physical universe.

Therefore, astronomers' quest for planets is not just motivated by their desire to find novel life forms and new regions of habitability. It is also driven by their interest in unraveling the secrets of the shadowy portions of the cosmos. For this reason, the extrasolar planet search is of importance to those studying the nature of the universe as a whole, as well as to those seeking new Earth-like domains.

Strange Rotations

At first glance, the dark matter mystery seems paradoxical. One might wonder how scientists know that most of the universe is invisible if they cannot observe these regions. How can they sense something that is telescopically undetectable?

As we have discussed, throughout the history of astronomy there have been a number of examples of objects discovered through means other than visual observation. The planet Neptune, for instance, was anticipated by Adams and Le Verrier on the basis of its gravitational influence several years before it was spotted in the sky by Galle. The tugging effect that Neptune had on the orbit of Uranus was enough to give its presence away.

Similarly, when Bessel discovered in 1844 that Sirius had an invisible companion, he found the new object through the gravitational perturbation that it created. By recording and analyzing the erratic behavior of Sirius, he computed the mass of its unseen partner. When in 1862 Clark telescopically observed the white dwarf, Sirius B, Bessel's calculations bore fruit. Clark's observations splendidly confirmed his predictions.

In recent years, the discoveries of planets around pulsars and giant worlds around Sun-like stars represent similar predictions based upon the evidence of gravity, rather than on direct visual indications. The work of planet-hunting astronomers such as Wolszczan, Mayor, Marcy, and Cochran (among others) shows that one needn't see something to know it is there.

The so-called "missing mass mystery" began with a similar kind of observation. In the early 1930s, Dutch astronomer Jan Oort discovered that the behavior of stars in the Milky Way cannot be entirely explained by the action of the gravitational force produced by their visible matter. Oort's research involved examining the dynamics of stars in the periphery of our galaxy. He was particularly interested in the upward and downward motions of these stars as they bob above and below the Milky Way's central disk.

Like ocean buoys in a storm, stars in our galaxy tend to rise or sink relative to the galactic plane (the imaginary surface that divides the Milky Way into top and bottom halves). The ascent or plunge of each star is caused by the gravitational tug due to all of the others. The amount of this pull depends on the total mass of the rest of the objects in the galaxy.

Exploiting this effect, Oort recorded stellar movements above and below the galactic disk and calculated how much gravitational force would be needed to maintain these motions. In particular, he measured the velocities of stars as functions of their height above or depth below the plane. From these measurements, he estimated the total mass of the Milky Way. He then compared this estimate to an appraisal of the quantity of visible matter in our galaxy—that is, the total mass of all of its stars. To his great surprise, he found that the total mass in the Milky Way required to generate observed stellar movements was at least three times the amount of the galaxy's visible material. The former quantity—Oort's estimate of the true mass of the galaxy—came to be known as the *Oort limit*.

The fact that the Oort limit was so much higher than the amount of observed matter in the Milky Way created something of a dilemma. Scientists wondered why so much material remained undetected. The difference between the Oort limit and the mass of all of the stars came to be known as the galaxy's "missing mass."

Back in the 1930s, few researchers were worried about this dilemma. Because relatively little was known at that time about the physical universe, they expected that further studies would eventually clear up the situation. However, soon additional findings about galactic movements served to deepen the missing mass mystery.

In 1933, shortly after Oort's star studies were completed, work by astronomer Fritz Zwicky provided even stronger evidence that much of the universe is made of invisible matter. Zwicky, who was born in Bulgaria in 1898, was a relative latecomer to astronomy. Three years after completing his Ph.D. in physics from the Federal Institute of Technology in Zurich, Switzerland, he moved to the United States. He worked there for several years in various fields of physics, including the areas of jet propulsion, crystal studies, and fluid dynamics before switching to astronomy.

Interestingly, Zwicky changed his area of research on the basis of a challenge. He once remarked to Robert Millikan, president of Caltech, that "any great physicist can change fields and make a name for himself within five years."[1] Millikan quickly issued an invitation for Zwicky to work at Caltech's recently established astronomy department, providing that he lived up to his word to make a name for himself. Sure enough, Zwicky accepted the offer and, within five years, became a powerful force in astronomy.

Zwicky's discovery of dark matter stemmed from an analysis of a grouping of galaxies called the Coma Cluster. First he computed the mass of the cluster by adding up how much material appeared to be in its constituent galaxies. Then he calculated the mass of the grouping a second way be estimating the gravitational attraction required for such a large system to be stable. Because the cluster is so spread out, he determined its mass to be quite high. Only an enormous amount of matter could provide the gravitational glue to keep it from flying apart into space. Amazingly, he computed the mass required to be 300 times the quantity he obtained in the first place, that is, by summing the amount of material in its galaxies. He surmised by this that the bulk of the Coma Cluster is invisible.

Astronomers' interest in the nature of dark matter rose in the late 1960s and early 1970s when Vera Rubin and her colleagues at the Carnegie Institution of Washington presented unmistakable proof of its existence. This evidence came by means of tracking the speeds of stars in distant galaxies. From the data gathered, the galaxies observed seemed to contain large quantities of unseen material.

Figure 31. Vera Rubin, of the Carnegie Institution of Washington, investigated dark matter in Andromeda. (Courtesy of the Carnegie Institution of Washington and Philip Bermingham Photography)

Rubin, who obtained her Ph.D. in astronomy from Georgetown University, began her examination of the dark matter question while investigating the Andromeda galaxy. Andromeda, our closest large neighbor, has a spiral shape similar to that of the Milky Way. Because of its proximity and resemblance to our own galaxy, it is an ideal subject of astronomical inquiry. Conclusions drawn from Andromeda's behavior generally apply to the Milky Way as well.

Rubin was particularly interested in Andromeda's rotation. Working at Carnegie along with researcher Kent Ford, she tracked the speeds of stars as they revolved around the center of that galaxy. They obtained their measurements by looking at their targeted objects' Doppler shifts. Recall that red spectral shifts indicate movement away from Earth, and blue, movement toward Earth. From the data gathered, Rubin and Ford graphed what is called a galactic rotation curve: stars' orbital velocities plotted versus their distances from galactic centers. By using a simple relationship they computed the mass distribution of Andromeda needed to produce such motions. They found out how much mass, spread out in which manner, would generate the gravity required to cause the stars to rotate around the galactic hub at the speeds they observed.

What they expected to find and what they did turn up were two different stories altogether. Spiral galaxies, such as Andromeda and the Milky Way, are shaped like fried eggs. Their centers, in which the bulk of their stars are located, bulge outward like yolks, while their outer reaches are thin. Because of their bulging shapes, it was long thought that most of the mass of spiral galaxies was concentrated within their central protuberances, and very little in their peripheries. For this reason, Rubin and Ford anticipated that Andromeda's rotation curve would indicate its mass to be heaped in its middle. Instead, they

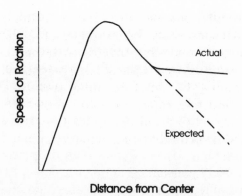

Figure 32. The visible mass of spiral galaxies, such as Andromeda, is concentrated in their centers. Therefore, one would expect that the rotational speeds of halo (peripheral) stars in these galaxies would be low. (Low mass produces low gravity and provides low speeds.) However, Vera Rubin showed that these speeds plateau, rather than drop off, with increasing radial distances.

were surprised that their plot revealed a much more uniform mass distribution.

To understand why the galactic rotation curve that they found was such a surprise, let us consider the velocity versus distance curve for our own Solar System. Instead of graphing stellar speeds, we plot planetary speeds. Applying Kepler's laws of planetary motion, we find that planets farther away from the Sun tend to move much slower than those closer to the Sun. This is because most of the Solar System's mass is concentrated in its center, namely in the Sun. There is very little matter in the Solar System's outer reaches. This highly concentrated mass distribution means that gravitational force drops off sharply with distance from the Sun. Thus, planetary velocities similarly show a sharp decrease with increasing distance. Pluto, for example, orbits the Sun much more slowly than Mercury does.

If Andromeda's mass were distributed according to the arrangement of its stars, then a similar decrease in velocities with greater distances from the center would be seen. In that case, its matter would be concentrated in its central bulge. Therefore gravitational forces—and stellar speeds—would drop off sharply with distance, in accordance with Kepler's laws of orbital motion. But Rubin and Ford found that the galactic rotation curve for Andromeda rose with distance and then reached a plateau. Far beyond its central bulge, stellar velocities continued to maintain high values. Rather than dropping off, they remained fairly constant.

Rubin and Ford's data indicates that the mass of Andromeda, rather than being concentrated in its yolk, is spread fairly evenly throughout the entire galaxy. There seems to be a substantial amount of matter in its halo: the peripheral region where few stars are present. The Carnegie researchers concluded that much of the material in Andromeda's halo must emit little or no discernable light. Thus, Andromeda is in some sense two galaxies: the visible spiral of stars and an invisible halo of unknown composition.

Rubin and Ford joined two of their colleagues, Norbert Thonnard and David Burstein, in investigating this phenomenon further. Was Andromeda's expansive mass distribution, brimming far beyond its stellar hub, just a fluke or did it represent something more profound? To decide this question, they selected 60 additional spiral galaxies for examination.

Not all galaxies are spirals. Only some that astronomers observe fall into that category. A significant number of others are elliptical in appearance, shaped like globs of putty. Still others are irregulars—their stellar material twisted into more elaborate kinds of patterns.

For pragmatic reasons, the Carnegie team chose to study spirals exclusively. The group realized that elliptical and irregular galaxies would be much harder to analyze. Only in spirals do the stars orbit in the same direction, like birds in a flock. In the other galactic types, stars move around in a more haphazard fashion. Therefore, it is easier to plot out the velocities of objects in spirals—and characterize these motions by galactic rotation curves—than to graph the movements of bodies belonging to other galactic varieties. Accordingly, the former, not the latter, kinds of stellar behavior were surveyed.

Rubin and her colleagues collected their data with the 13-foot-diameter telescopes at Kitt Peak in Arizona and at the Cerro Tololo Observatory in Chile. Because the spirals they examined were farther away than Andromeda, and consequently appeared smaller, they found they could image entire galaxies at once. With Andromeda, they had been able to look at only small sections at a time. As a result, they scrutinized the 60 new spirals at a much faster pace than they had examined Andromeda.

In short order, they obtained galactic rotation curves for each of their subjects. Not surprisingly, some of the spirals were found to rotate faster than others. The more tightly bound galaxies whirled at the quickest rate; the looser ones spun at the slowest. This could be explained by the *principle of angular momentum*. Accoridng to this law, when a body draws itself in, it speeds up, and when it extends itself, it slows down. Because spirals are supposed to have similar origins, it is natural that those that spread out tend to slacken their pace.

Not everything the team observed could be justified by the then-known laws of physics and astronomy. Like television screens set to the same channel, lined up in a department store, the rotation curves each galaxy generated showed the same basic picture. One by one, the group plotted the data from each galaxy and were amazed to see a graph virtually identical to the others that they analyzed. In each case, in charting stellar motions versus distances, they found that velocities rose with distance throughout the inner galactic zone and then came to a plateau in the peripheral region. Something was driving matter to move in the outer reaches of these spirals much faster than they had expected.

As in the case of Andromeda, Rubin and her co-workers surmised that there was more to these galaxies than meets the eye. Now that they had examined such a large sample, they were even more certain that much of the material in the haloes of galaxies is dark. Considerable quantities of invisible substances seemed to lurk in the outer reaches of spirals, propelling, with their combined gravitational force, peripheral objects in these galaxies to move much faster than they would have otherwise. Either that, or the laws of gravitational physics were dead wrong.

The renowned medieval English philosopher William of Occam preached that the most rational solution to a puzzle involves the fewest possible assumptions. Newton's principle of gravitational attraction, as modified by Einstein, has served us well for centuries in modeling a vast range of phenomena, from the dropping of apples from trees to the sweeping behavior of comets. Newton's work is so elegant and has such profound universal predictive powers that it would seem folly to scrap it for idle reasons. It is far simpler to assume that there is much unseen matter in galaxies. Therefore, for astrophysical theorists pondering the Carnegie results, "Occam's razor" seemed to whittle away fanciful notions that Newton's theory should be modified and leave intact the more convincing premise that much of the mass of galactic haloes could not be detected.

In 1973, Jeremy Ostriker and Jim Peebles of Princeton University completed a computer study that lent even further credence to the dark matter hypothesis. Using state-of-the-art simulation programs, they showed that spiral galaxies would only be stable if they were engulfed in dark haloes. They calculated that each invisible cloud would need to have about 10 times the observed mass of the galaxy that it surrounded, or else the spiral would eventually become unfurled. Because strong evidence indicates that spiral galaxies do not unwind in such a manner, they concluded that dark haloes must exist to provide galactic stability. They speculated that this unseen halo material was composed of stars too dim to be recorded.

After Ostriker and Peebles's results were published in the prestigious *Astrophysical Journal*, the search for dark matter in the cosmos began in earnest. With two established researchers with powerful reputations stating that the bulk of galactic material cannot be seen, their paper was received with the utmost seriousness. The dam was broken, and a tidal wave of speculation about the "missing mass mystery" burst forth.

In 1991, the enigmatic substance that appeared to occupy the outskirts of galaxies acquired a colorful name. Upon the suggestion of astronomer Kim Griest of the University of California at San Diego, MACHOs (an acronym for MAssive Compact Halo Objects) became the common term for dense bodies seemingly present in peripheral galactic regions. (For convenience, we shall use this term even when referring to earlier studies.)

Conjecture abounded as to the possible nature of these objects. Many astronomers surmised that they were likely some combination of red dwarfs, brown dwarfs, dim white dwarfs, and extrasolar planets. Additional possibilities included neutron stars, black holes, and other types of exotic objects. The only way to determine the true composition of these bodies was to chart them and analyze their makeup. But how, astronomers wondered, might one see the invisible?

Spacetime's Wrinkled Fabric

In the 1980s, dozens of researchers devoted themselves to trying to resolve the dark matter question. At first, the problem seemed quite formidable: if massive halo objects give off no discernable radiation, how indeed might their presence be made known and their nature be understood? How might astronomers decide, for instance, how much of this halo matter represents extrasolar planets?

Soon, many realized that the natural solution to this dilemma was measuring the gravitational interactions of MACHOs with other objects. They sought ways of sensing the gravitational pull of dark matter, thereby mapping out its density and location. In this way they hoped to be able to decide the composition of this material; what fraction of it is red dwarfs, brown dwarfs, white dwarfs, black holes, planets, and other objects?

The method that emerged for performing this feat was the technique of gravitational lensing. Gravitational lensing is an effect in which material situated between Earth and a distant body affects the appearance of the faraway object. This method derives from the fact that, according to Einstein's principles of general relativity, heavy matter can bend light, and even focus it like a lens. Unseen material, assuming that it exerts a gravitational influence, is no exception to this rule. Such matter might consequently be probed by means of its lensing properties.

Remarkably, the use of gravitational lensing to probe dark matter was first suggested as early as 1937. Fritz Zwicky in a brief note to the *Physical Review* wrote that galaxies could be better "weighed" if they were found to act as lenses. But Zwicky's idea for mapping out dark matter languished in obscurity for almost half a century, until it was finally taken up again in the 1980s. Zwicky died in 1974, before he could see his notion put to the test by the scientific community.

Before we examine how MACHOs might serve as gravitational lenses and therefore be analyzed by their affect on light, let us first discuss the basic notions of general relativity. This theory shows us how any massive object in the cosmos might be examined through its gravitational influence.

The theory of general relativity, considered Albert Einstein's masterpiece, was published in 1915. Like Newton's work, it is a mathematical scheme for describing how objects are affected by the gravitational influences of other bodies. Both models provide means of predicting the orbital behavior of celestial bodies, such as planets and stars. Whereas Einstein's theory is believed to be universally accurate, Newton's only applies in the case of weak-to-moderate gravity. Einstein's work thus supplants and expands Newton's set of principles.

General relativity addresses the weakest element in Newton's theory of gravitation: the idea of action at a distance. According to Newton, one object attracts another by means of a force exerted between them. This long-range attraction is proportional to the product of their masses and inversely proportional to the distance between them. It is unclear, though, how this force is transmitted if the objects aren't physically touching each other. Newton simply assumed that objects can affect each other no matter how far apart they are.

In contrast, Einstein's theory makes no such assumption. Gravity, according to his work, is transmitted by means of fluctuations in the fabric of space itself. To be more precise, Einstein's model groups space and time together into a single entity called spacetime. Gravity, then, represents ripples in the structure of spacetime that are caused by the presence of mass in any particular region. These ripples, in turn, affect the movements of nearby objects. Thus gravitation, rather than operating over a distance, acts locally (at a point) with spacetime as an intermediary.

The relationship between the mass distribution and geometry of a spacetime sector can be mathematically characterized by the so-called Einstein equations. According to these relations, the material content of a region precisely governs its curvature. This distortion, in turn, determines the paths taken by all objects in that zone. Theorists Charles Misner, Kip Thorne, and John Wheeler have summarized this well in a book on the subject: "Space acts on matter, telling it how to move. In turn, matter reacts back on space telling it how to curve."[2]

The difference between the two models—Newtonian and Einsteinian—might best be understood by means of analogy. Imagine a wizard with magical powers, who has the ability to make others instantly feel hot. All he needs to do is think the word "hot" and those around him feel as if they are burning up. In this manner, the wizard applies action at a distance, in a way analogous to Newtonian forces.

Now picture a second wizard, who happens to be a fraud. The fake wizard wears a ring with a tiny thermostat that is connected to a powerful space heater. Whenever he wants to impress those around him, he secretly pushes the button on his ring, causing the heater to turn on. Soon everyone in his vicinity experiences a scorching sensation. In this case, the air in the room acts as an intermediary, transmitting the wizard's power through convection. In analogous fashion, spacetime conducts gravitational action by means of fluctuations—created by one object or set of objects and conveyed to others.

Briefly, here's how the principles of general relativity

work. Spacetime, according to Einstein, is said to be "flat" in regions where no matter is present. Flatness here is something of a metaphor. Einstein's spacetime is a four dimensional entity, composed of the three dimensions of space and one of time. When this composite is not distorted it is referred to as flat.

Drawing on a two dimensional analogy to aid us in imagining this, we might picture spacetime as a rubber sheet, stretched out to its fullest extent. Clearly, if no massive objects rest on it, the rubber surface would neither sag nor bulge; it should remain completely level. Thus we say that, in the absence of matter, the sheet (and spacetime, by analogy) is flat.

Now imagine rolling a marble along the sheet. Because its surface is planar, the marble must follow a straight trajectory as it rolls across it. Only in this manner would the marble be taking the most direct route. Similarly, a celestial object, such as a star or chunk of rock, traveling through flat space must move along a direct, linear path. Even light particles must follow the shortest possible routes through such regions. That is why in empty space, sufficiently far from massive objects, light rays always consist of perfectly straight lines.

Next picture a heavy bowling ball being placed on the sheet. The placement of this ball is analogous to the positioning of a massive body, such as a pulsar or large planet, in a particular region of the cosmos. Just as the bowling ball would cause the sheet to sag downward, the celestial object would influence its vicinity in spacetime to become distorted.

With the rubber sheet now distorted, any marbles rolled across it would tend to follow curved, rather than straight paths. Moving along the shortest possible trajectories, they might even end up circling around the ball. Similarly, astral bodies traveling through warped regions of space tend to follow curved paths. Because, for example, in the Solar System, the spacetime region near the Sun is curved, planets and comets follow elliptical orbits around it.

In a critical departure from classical (Newtonian) physics, Einstein's model states that even light rays, in the presence of massive bodies, must follow curved paths. The paths of light particles are altered for the same reason that the trajectories of other objects are deviated; light pursues the shortest route possible. In the presence of matter, the most efficient path is curved. Thus light's trajectory is necessarily bent by gravitational distortion.

In comparison, Newtonian theory (coupled with what is now known about light particles) mandates that gravity should not have any influence on light. Classical physics states because light particles are massless, they should not be affected by the presence of massive bodies in their vicinity. Therefore, even in the midst of strong gravitational fields, such as those near heavy objects, light must still travel along straight lines. (Newton himself believed that light would bend slightly, though, because he erroneously thought that it had rest mass.)

Soon after the theory of general relativity was published, it was put to the test. British astronomer Sir Arthur

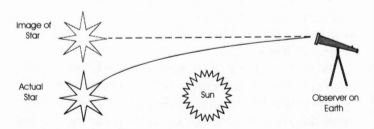

Figure 33. Einstein showed that the mass of the Sun can serve to deflect stellar light, shifting the observed position of a star.

Eddington took up a suggestion by Einstein that the bending of starlight by the Sun would be a sure way of testing his theory. Einstein had pointed out in his 1915 paper that a total eclipse of the Sun would be a perfect occasion to measure relativistic distortion using the light from distant stars. Normally the Sun's brilliant glare would prevent observation of stellar positions near its edge. But during an eclipse, its bright disk would be covered, and the locations of stars around its halo could be measured.

Eddington could do little to test Einstein's hypothesis while World War I still raged. As an Englishman of that time, it was difficult to be a vocal advocate for investigating the scientific views of a German. Besides, the ideal conditions for testing had not yet arisen.

However, both Eddington and Einstein were men of peace and managed to keep up communication in spite of the bloody conflict between their two countries. For a time, Eddington was one of only a handful of people in the world who understood and appreciated general relativity. As well as he could, he articulated the importance of Einstein's ideas to his scientific colleagues, and, later, to the general public.

Once the war ended, Eddington saw a great opportunity to prove that his correspondent was correct. A solar eclipse was coming up in which the Sun would be ideally positioned in front of the bright Hyades star group. The Sun's location would be such that the rays from these stars would curve around it on its way to Earth. Therefore, their positions would appear shifted in the sky by a predictable amount. And because the solar disk would be obscured, the altered stellar coordinates could readily be compared to their values when the Sun wasn't in the way. In short, he realized that it would be foolish to pass up such a bold chance to confirm one of the key forecasts of general relativity.

Along with several other British researchers, Eddington began preparations for expeditions to record stellar positions during the May 1919 solar eclipse. The best places to view the darkened morning sky were determined to be in the tropics. So A. C. D. Cromellin and C. R. Davidson of the Royal Observatory at Greenwich sailed out to northern Brazil, and Eddington, along with colleague E. T. Cottingham, headed over to Principe Island in the Gulf of Guinea. Each team set up its equipment and waited for the total eclipse to begin.

Eddington's group had rather bad luck. A torrential downpour filled the skies of Principe right before the long-awaited event was to begin. By the time the eclipse had reached its peak, the rainstorm had slackened off enough for the team to expose a limited number of photographic plates. Unfortunately, only two of the photographs ended up being suitable for measurement.

Luckily, the Brazilian expedition had a much better experience. With clear skies at the time of totality, they were able to record 26 pictures of the Hyades. Although only seven of these ended up being usable, their date proved reasonably precise. In July, weeks after the eclipse was over, they returned to the site and exposed some more plates of the same group of stars. Then, by comparing the two sets of images, they measured the shifting of stellar images due to the intervening mass of the Sun.

Eddington's team and the other group combined their collected data and calculated how close their results came to Einstein's predictions. Remarkably, their experimental conclusions fell within 15 percent of general relativistic forecasts. Considering the sparsity of their data and their poor working conditions, the accuracy that they achieved was incredible.

Within five years of its formulation, Einstein's theory had been splendidly confirmed. Ironically, it was an elite group of his countrymen who wrested stodgy Newton, a

Cambridge don, from his hallowed throne of authority in physics, and replaced him in that position with a rumpled German Jew who used to earn his living in a Swiss patent office. Such was the revolutionary nature of the postwar years in Europe.

Gravity's Rainbow

Einstein realized that one consequence of the relativistic bending of light was that massive objects could serve as lenses. Under the right circumstances, if a large clump of material happened to be situated directly between a star and Earth, then the matter could focus that star's light rays into a pattern of images. He did not pursue the idea further, however, because he thought that this effect would require too much of a coincidence. He estimated that the chances would be extremely low for two stars, or a star and another object, to be lined up in the exact manner needed for gravitational lensing to be noticeable.

When Zwicky considered the matter in 1937, he had a better idea. If the gravitational lens being observed were a galaxy, rather than a star, its mass would be billions of times greater. Consequently, Zwicky argued, a galactic lens would bend and focus light to a much sharper extent than would a stellar lens. Such effects would be pronounced enough to be readily observed. He urged fellow astronomers to scan the heavens for evidence of gravitational lensing by galaxies, which he bet they could find without much difficulty. But it was decades later before anyone heeded his rallying call.

In the 1960s, theoretical physicists dug up the almost forgotten papers by Einstein and Zwicky and began to ponder seriously how the image of a gravitational lens would appear in space. What would happen, they wondered, if a distant galaxy's light were focused by the mass of another galaxy that happened to lie between it and Earth? Calculations showed that terrestrial observers would see one of several kinds of patterns, depending on the relative positions of the two galaxies.

If the galaxies were lined up exactly, pointing like an arrow toward Earth, then the image of the more distant object would appear in the skies as a ring of light. Astronomers call this type of image an Einstein ring. This would happen because the light rays from the more distant galaxy would bend symmetrically around the intervening object before they reached Earth. Tracing these rays backward into space from where they seemed to emanate, terrestrial onlookers would see them as forming a circle.

If, on the other hand, one of the two galaxies was a bit off center relative to the line defined by the other one and Earth, then the pattern created through lensing would no longer be symmetric. Rather, the light rays from the lensed galaxy would bend in a lopsided manner around the lensing body. The irregularity of such distortion would break up the image into several distinct components. As a result, the pattern seen by an astronomer gazing at the region with his telescope would consist of two or more blotches or crescents, rather than a unified ring.

The action of a gravitational lens might be pictured by imagining a team of skiers skiing down a slope at a steady speed. Suppose an observer were watching the skiers from the bottom of the hill trying to guess if they all started together. If he saw all of them heading straight down in a pack, each in synchrony with the others, he might very well guess that they began their descent as a group. On the other hand, if he noticed that one set of skiers was approaching him diagonally on his right, and another, diagonally on his left, he might suppose that there were two original clusters, each leaving from a separate point. He would thus conclude that not all of the skiers started in unison from the same position. Similarly, astronomers observing two blotches of light in the sky might assume that they originated from two different sources.

But the ski-watcher might be wrong about his conclusion. Suppose the skiers did start out as a bunch, but got diverted at some point along their path. A large boulder could have blocked them from coming straight down, forcing them to swerve around it. Perhaps, in avoiding the rock, half of the original group swerved to the right and the other half to the left. Then, as each fraction headed toward the finishing point, they tried to merge into a unit again. The ski-watcher, in seeing this final step, but not its precedents, could easily have jumped to the wrong conclusion. In similar fashion, astronomers, seeing light after it has swerved around a massive intervening object, might easily reach the wrong conclusion that the radiation came from two different sources rather than one.

It wasn't until the late 1970s—a time when telescopy grew greatly in precision—that a pattern produced by a gravitational lens was first observed. This astronomical breakthrough was announced in the May 1979 issue of *Nature* by Dennis Walsh of the Jodrell Bank Observatory, Robert Carswell of Cambridge University, and Ray Weymann of the University of Arizona. Using the 7-foot telescope at Kitt Peak Observatory, they found a strange pair of quasars near the Big Dipper that possessed an extraordinary array of similar features. Quasars are extremely distant objects that, though relatively small, shine with the brilliance of galaxies.

The two quasars, called 0957 + 561A and 0957 + 561B, were soon nicknamed "the twins." Though 6 seconds of arc apart in the sky (about 300 times more narrow than the angle that the Moon subtends), their spectral lines were identical in all respects, except that quasar A

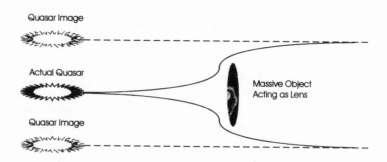

Figure 34. If an extensive massive object passes between Earth and a distant quasar, it might serve as a gravitational lens, breaking up the quasar's light into multiple images.

was brighter. Moreover, each set of spectra was Doppler shifted toward the red by the same amount. This indicated that the quasars were equally distant from Earth—each billions of light years away. Coincidence seemed to pile up on coincidence until the Kitt Peak team became convinced that they must be seeing double. They realized they were seeing a single quasar through a kind of cosmic fun house mirror; quasar A and quasar B were two reflections of the same object. A massive elliptical galaxy, weighing more than one trillion suns, was breaking up the light from 0957 + 561 and recasting it into two distinct images. Because of this gravitational lensing, the group observed these images in different parts of the sky and mistook them, at first, to be separate bodies.

The discovery at Kitt Peak opened the door to numerous additional sightings of lensed quasars and galaxies. As astronomers became more and more adept at recognizing images produced through gravitational lenses, Zwicky's dream of weighing galaxies with such means seemed increasingly realistic. Groups began to plan, using gravitational lensing to help solve the dark matter mystery. They hoped that analysis of images lensed by MACHOs would help reveal their identity.

MACHO Detectives

Three large projects were formed in the early 1990s to scout out and identify the dark matter believed to be present in galactic haloes: a French group (EROS); a Polish team (OGLE); and an American–Australian collaboration (MACHO). All three groups sought examples of gravitational lensing by massive objects in the Milky Way's halo of distant stars in the Large Magellanic Cloud. (The Large Magellanic Cloud is a collection of stars and gases that acts as a satellite system to the Milky Way.) They hoped to record and analyze incidents in which invisible bodies in our galaxy distorted starlight from our galactic neighbor.

The EROS team began its lensing studies in 1990. In French, EROS means "Experience pour la Recherche d'Objets sombres," which loosely translates to "Dark Object Research Experiment." The main objective of the group is the research and the study of dark stellar bodies, especially brown dwarfs, that lie in the outer reaches of the Milky Way. Conducting its work at the European Southern Observatory in La Silla, Chile, the team includes Eric Aubourg of Saclay, Jean-Philippe Beaulieu of the Institut d'Astrophysique de Paris, and Reza Ansari of the Linear Accelerator Laboratory at Orsay, among many others.

The Polish-led group OGLE, whose name signifies "Optical Gravitational Lensing Experiment," began its dark matter research in 1992. The collaboration is headed by Andrzej Udalski and Michal Szymanski of Warsaw University Observatory. Besides looking for lensing by MACHOs of stars in the Large Magellanic Cloud, the group also scans for lensing of the galactic bulge stars. Each night, using the 3-foot Swope telescope at the Las Campanas Observatory in Chile, the team monitors about two million stars, looking for lensing events.

The third group, the MACHO collaboration, represents the largest galactic dark matter research project. The team is headed by Charles Alcock of Lawrence Livermore National Laboratory, along with David Bennett of Lawrence Livermore and Kim Griest of the University of California at San Diego. Starting in June 1992, they have conducted their searches using the 5-foot diameter telescope at the Mount Stromlo Observatory near Canberra, Australia.

The technique that these groups have used is often referred to as "microlensing." In microlensing, as opposed to conventional gravitational lensing, a star's rays are distorted by an object of planetary or brown dwarf size, rather than of galactic proportions. Because it involves such diminutive objects, microlensing searching represents a method that requires great precision. To achieve this, CCD (charge-coupled diode) cameras are placed in front of the telescopes being employed, and special computer programs are used to reconstruct images from the data.

In 1993, the work of these teams came to fruition when they detected a MACHO for the first time. Each group sighted a massive halo object that briefly passed between the Earth and a red giant from the Large Magellanic Cloud. As the MACHO moved by, it momentarily concentrated the light from the star, amplifying the signals recorded by the three collaborations. Each recorded it as an increase, then decrease, in the intensity of the red giant's image. The entire process was measured as taking place over a period of 33 days. Computer analysis ruled out other possible sources of the signal; its pattern could only characterize a MACHO. This pulse was further evaluated to determine the mass and extent of the invisible object. Finally, satisfied that their judgment was correct, all three teams announced success.

Within the next few years, the MACHO, EROS, and OGLE collaborations continued to search for dark matter in the galaxy. The MACHO group seemed to have the best luck, concentrating its efforts on unveiling lensed images of stars in the Large Magellanic Cloud. By the end of 1995, they had found seven additional examples of massive objects in the Milky Way's halo. They reported their promising results at the January 1996 meeting of the American Astronomical Society.

During their presentation, members of the team drew several important conclusions about their findings. First of all, they speculated about the nature of the MACHOs they had discovered. They found these objects to be small, ranging from one-tenth solar mass to full solar mass. The researchers conjectured that, because of the objects' meager size and manifest dimness, most were likely white dwarfs, with a few red and brown dwarfs among the lot.

From the quantity and types of MACHOs detected, the members drew further conclusions about the prevalence of these objects in our galaxy. Churning their results through detailed statistical analysis, they estimated that MACHOs comprise close to 20 percent of all of the halo dark matter in the Milky Way. Although the identity of the other 80 percent remained in dispute, the superb results of the Australian-American team, as well as those of the French and Polish groups, brought the missing mass problem closer to resolution than ever before.

PLANET Sleuths

One might wonder where extrasolar planets fit into this picture. Certainly planetary bodies form a fraction of the unseen material in space. It is an open question, however, as to how large is this fraction.

When the expansive MACHO search projects began in the early 1990s, one of the explicit goals of these missions was to hunt for new worlds. Theorists realized early on that the technique of gravitational microlensing, so useful in the discovery of dim stellar dwarfs, would also be sensitive enough to detect planets.

In 1991, professors Shunde Mao and Bohdan Paczynski of the Princeton University Observatory laid the firm theoretical ground for using microlensing to seek out extrasolar worlds. In an influential paper, they noted a curious effect of solar-size gravitational lenses: their light distorting region has close to the same size as the orbit of a Jupiter-size planet. Because of this coincidence of scales, planetary systems acting as lenses would produce characteristic types of images. Mao and Paczynski predicted that this effect could be observed by astronomers in the light intensity curve (magnification versus time) of distant stars if the starlight were momentarily lensed by Jupiter-size planets. That is, the optical equipment would record a special "blip" as the "jupiter" passed between the star being observed and Earth.

The Princeton astronomers felt that the ideal conditions for observing microlensing planetary systems would occur when they passed between galactic bulge stars and the Earth. This has to do with our world's position relative to the center of the galaxy. Earth is located on one of the Milky Way's spiral arms. Undoubtedly, there are numerous planetary systems that lie between our region and the central part of the galaxy. To observe these by means of their lensing properties, one must peer through them at the light from more distant objects, namely the stellar giants that occupy the galactic bulge.

This situation is like that of a farsighted woman, wishing to test out a new pair of reading spectacles. To see how well they worked, she wouldn't want to gaze through the glasses at a road sign a mile away, or stare through them at the tip of her nose. Rather she would likely wish to peer at something at arm's length, such as a book, and check how she can make out the words.

In similar fashion, to examine the properties of a gravitational lens, one would like to chose targets that are a moderate distance away. For planetary systems within our galactic arm, the large stars inhabiting the core of the Milky Way possess the appropriate distances. These stellar giants in the central region of the galaxy are far enough away to be lensed by intermediate planetary systems, yet luminous enough to be readily studied.

Mao and Paczynski estimated that about 10 percent of all lensing incidents involving galactic bulge stars will indicate that the lens constitutes a multiple system. While some of these complex lenses may represent stellar pairs, a significant fraction may comprise planets in orbit around their suns. The Princeton researchers speculated that even Earth-size planets might be found using this technique. They concluded their study by urging a methodical search for microlensing of the galactic bulge.

In 1995, an international microlensing monitoring group was formed called PLANET (Probing Lensing Anomalies with a world-wide NETwork). The main goal of this collaboration is to analyze anomalous microlensing events with the hope of detecting extrasolar planetary bodies. The team includes scientists from South Africa, Australia, Germany, and the Netherlands. Their research is conducted with three widely separated telescopes in the Southern Hemisphere: the Dutch-ESO instrument at La Silla, Chile; the Perth Observatory in Bickley, Australia; and the South African Astronomical Observatory in Sutherland, South Africa. Because these telescopes are so far apart, it has never been the case that all of them experience bad weather at once. Therefore, during the prime observation season—when the center of the galaxy is visible—there is always at least one telescope scanning the skies for interesting phenomena.

PLANET decides where to conduct its searches based on initial data that it obtains from one of the three major dark matter groups: MACHO, OGLE, and EROS. Each of these groups supplies PLANET (and anyone else) with periodic microlensing alerts and frequent reports on unusual patterns, whenever the galactic bulge is visible. Based on this information, the PLANET astronomers elect to study in detail a certain fraction of these events, hoping to observe the characteristic signature of a light-distorting extrasolar world.

In the opening year of its survey, PLANET reported its first sighting of a possible planetary system. While interpreting and extending a study initiated by MACHO, PLANET astronomers discovered the signal of a sun and orbiting body passing in front of a galactic bulge star. As the radiation from the bulge star swerved around the intervening system, it grew concentrated by the system's gravity. This was observed by the members of PLANET as a sharp peak in the star's luminosity (light output) versus time curve.

The PLANET collaborators continue to scan for examples of planetary microlensing. They have been joined in their quest by MOA (MACHO Observations in Astrophysics), an independent team based in New Zealand. Given the power of their technique, and the limitations of the Doppler (radial velocity) method in sensing planets much smaller than Jupiter, microlensing groups may well be the first to find an Earth-size world outside of the Solar System.

The Other Side of Matter

MACHOs, including stellar dwarfs and planets, account for about 20 percent of the Milky Way's invisible halo material. One might wonder what comprises the remainder of the dark matter in the galaxy. Furthermore, experiments have shown that intergalactic space contains unseen mass as well. The void regions of space are not as empty as once believed. What, then, is the mysterious substance or mix of substances that cloaks the cosmos?

There has been much speculation about the nature of the non-MACHO portion of the universe's dark matter. In contrast to stocky, solid MACHOs, astronomers refer to the more etherial component of dark matter by the term "WIMPs." This acronym, which stands for "Weakly Inter-acting Massive Particles," refers to the propensity of WIMPs to interact with ordinary material only by means of the weak and gravitational forces and never through electromagnetic coupling.

The weak interaction, one of the four fundamental forces of nature, represents the least potent way in which particles might associate. Objects that interact weakly also do so rarely. Because of this infrequency of contact, particles subject to the weak force can travel for very long distances without colliding with others. And, as long as they don't feel the electromagnetic force, they must re-main unseen through their journeys. This invisibility stems from the fact that light particles are produced ex-clusively through electromagnetic processes. Therefore WIMPs would not generate luminous radiation as they scurried through the universe. Yet, because they would have mass, their presence would be felt by means of their gravitational attraction. For this reason, WIMPs would comprise suitable constituents of the murky material that has so far eluded telescopic detection.

In recent years, theorists have suggested hundreds of possibilities for what WIMPs (and other minute forms of dark matter) would entail. Assorted contenders include supersymmetric companions to elementary particles, as well as massive neutrinos, peculiar objects called axions, and elusive shadow matter. Although the details of this menagerie fall well beyond the scope of this book, I will briefly comment on these.

Supersymmetry is a hypothetical property that some theorists believe describes the early universe's unified state. It is a theoretical answer to a question that has puzzled modern physicists: why is it the case that for one class of elementary particles, called fermions, no two objects in the same position can have exactly the same properties (charge, energy level, spin, etc.), whereas for another class, called bosons, an indefinite number of identical particles can congregate together? While electrons and protons (atomic constituents), falling into the former category, obey one set of statistical laws, photons (light particles), belonging to the latter group, obey a completely different type of statistics.

Those who advocate the supersymmetry hypothesis maintain that the young cosmos, when it was extremely hot, possessed only one class of particle. As the universe cooled down, the initial state of supersymmetry became broken, and the fermionic and bosonic categories emerged.

One consequence of supersymmetry theory is that every fermion is said to have a massive bosonic companion and vice versa. These companions have properties similar to ordinary particles except for their masses and spin states. The hypothetical mates of electrons are called selectrons, for example, and those of photons, photinos.

Supersymmetric companions have yet to be found in nature. Yet if they do exist, their presence would answer many questions. Their combined mass would form a significant chunk of the WIMP (non-MACHO) segment of dark matter. For this reason, many high energy experimental physicists search fervently for these objects, hoping, with their discovery, to solve one of the central mysteries of the universe.

Other possible components of dark matter are massive neutrinos. Neutrinos, among the lightest and most common of elementary particles, are formed during radioactive decay and other types of subatomic processes. Myriads of neutrinos are expelled by the Sun as it fuses hydrogen. Although they have traditionally been thought to be massless, there is some evidence that they might have some very small mass. If neutrinos indeed have mass, however minuscule, then there are so many of them that they would form a substantial portion of the dark matter in the cosmos.

Axions, the favorite candidate of leading dark matter researcher Michael Turner of the University of Chicago, are yet another possible component. These are hypothetical particles suggested by Frank Wilczek of Princeton to solve certain theoretical problems in field theory. Though inconceivably small, they would be so common (a billion per cubic inch of space) that they would add up to a hefty chunk of the unseen cosmic mass.

Probably the most outlandish dark matter theory is the notion of shadow matter—another one-time pet candidate of Turner. Shadow matter is a highly speculative type of material that interacts with ordinary substances solely through its gravitational force. According to conjecture, it cannot feel the weak, electromagnetic, or strong interactions—the other three known forces of nature. Therefore, it might only be seen through its gravity.

If much of space were made of shadow matter, then the universe would be vastly more mysterious than even imaginative science fiction scenarios have suggested. Vast regions of the cosmos would be cloaked in garments of invisibility. Because shadow matter would be almost impossible to sense, much of the universe would be virtually incomprehensible.

Perhaps, somewhere in the nether reaches of space, there are stars, galaxies, and even planetary systems made out of this mysterious substance. Because they would emit no visible radiation, veteran astronomical teams, such as Marcy and Butler's group, could never detect such suns or planetary bodies. Neither Doppler detection nor astrometry would reveal these ghostly worlds. If such bodies existed, they would probably be detected solely through their gravitational lensing effects.

In theory, whole invisible civilizations might exist, replete with beings composed of shadow matter. Curiously, we could never hope to communicate with such aliens through conventional means, such as radio signals. Because they would be insensitive to electromagnetic radiation, such messages pass right through them undetected. Similarly, communications that they might send would pass right through us unhindered. Possibly the only way we might talk with such creatures would be by means of gravitational waves—signals that both societies could conceivably detect. We would need to develop special translating devices in which terrestrial messages could be encoded as gravitational pulses. Unfortunately, modulated gravitational waves would be extremely difficult to produce and detect, so generating and deciphering such communication would be tedious at best.

At present there is absolutely no physical evidence of shadow matter, thus it would be foolish to devote precious scientific resources to an attempt to find shadow worlds. Even Turner devotes little time to this notion, which he views as fanciful. Until experimentalists turn up proof that such a substance exists, talk of shadow creatures had best be uttered well beyond the hallowed meeting rooms of professional scientific conferences.

Our Cosmic Fate

One of the reasons scientists are so fascinated by the dark matter dilemma is that it has a strong bearing on the question of the fate of the universe. Whenever the number of stars, planets, and other assorted material in space is tallied, they examine this quantity closely to see what it indicates about the density of the universe. They calculate how much mass on average fills every square inch of space. Then, they use the density of the cosmos to predict its ultimate destiny.

If you are reading these words on a hot summer day, it may be difficult to contemplate the cruel fact that the world is finite. Sitting on the beach, soaking in the fiery energy of the Sun, you might find it unthinkable that someday the overhead source of warmth that steers the surf and bakes the sand will cease to shine. Yet astronomers tell us that in billions of years, the solar furnace will exhaust its fuel and metamorphose into a red giant, then into a cold white dwarf, and finally into useless interstellar debris.

Or perhaps, ages before the Sun burns out, Earth will become uninhabitable for other reasons. Nuclear, biological, or chemical warfare may someday render our planet unfit for humankind. Even if we manage to avoid these global catastrophes, rampant pollution or unchecked population growth may eventually destroy the fragile ecology of our world.

One might hope that by then, thanks to Mayor, Marcy, and their successors, the human race will have found thousands of new planets orbiting nearby suns in which it might resettle itself. Perhaps by the time our own Sun expires and renders our system lifeless (or if, before then, our world becomes uninhabitable due to other factors) we will have long abandoned Earth and emigrated to one or more of the newly discovered worlds.

Gradually over the eons, though, each of the stars in our galaxy will exhaust its fuel and cease to shine. Our species may evolve into a race of celestial wanderers, roaming from one planetary system to another as each becomes extinct. Eventually, we may have to abandon the Milky Way itself and make the trek to other galaxies. Finally, after the last star in the ultimate galaxy has expended its final drop of fuel, the human race—if it still exists—will likely have to call it quits.

Modern cosmology tells us, however, that this drawn-out scenario of gradual universal extinction is only one possible fate of the cosmos. If gravitational theorists are correct, there is a chance that space will close in on itself a long time before all its usable energy has been exhausted. Applying general relativity to the universe as a whole, they have calculated a quantity called the critical density of space. If the actual density of matter and energy is greater than its critical value, then they predict the universe will eventually collapse like a deflated balloon.

If one does a careful count of all of the visible matter in the sky—from asteroids to quasars—then one finds that there is not enough mass for the universe to collapse. Based on the amount of luminous material in space, its actual density is far less than its critical density. If what we telescopically observed were all that existed, then we wouldn't have to concern ourselves with the possibility of a cosmos that might eventually contract.

Clearly, what we see with our viewing instruments isn't everything that's out there. If we've learned one thing from late 20th century astrophysics, it is that most of the cosmos gives off little or no discernible radiation. What is still unknown, though, is precisely what percentage of the cosmos is dark. And it is that figure, indicating the density of all of the matter and energy in space, luminous and nonluminous, that would cue us in to our fate. This is the important task assigned to those searching for dark matter—to determine exactly this consequential value, and thereby learn the destiny of us all.

Regardless of the ultimate fate of the cosmos, our civilization should make the best of the time ahead of it.

In the millennia to come (we hope), we must extend the boundaries of the human frontier and strive to explore as much of space as technology allows us. The nascent steps toward that goal have already been taken. In the past few decades, we have set foot on the Moon and sent unmanned probes to other planets within the Solar System. Astronauts and cosmonauts have shown that prolonged stays within space capsules are physiologically possible. And now researchers have proven that planets exist beyond the Sun's nine-world domain.

Like a child catching a glimpse of a pile of presents beneath a Christmas tree, we sense the miraculous nature of what more is to come. Each planetary system in space is a package waiting to be unwrapped, its wonders and delights begging to be explored. Before time draws to a close, we must reveal the gifts that nature has provided.

The next step in our exploration of new worlds in the cosmos is to seek out Earth-like planets that orbit stars similar to the Sun. Once these are found, we need to refine ways of determining whether or not they harbor life. If they do, we might consider attempting contact (which would be much easier with ordinary planets than with shadow worlds) and, once the technology became possible, send out missions to those worlds. And then, after the first interstellar cruiser has been launched, the true human adventure in space will begin.

Chapter 7
SIGNS OF LIFE

> *But, again, the worlds also are infinite, whether they resemble this one of ours or whether they are different from it. For, as the atoms are, as to their number, infinite ... there is ... no fact inconsistent with an infinity of worlds.*

> EPICURUS, *Letter to Herodotus*

Questing Other Earths

Since ancient times, philosophers, scientists, and theologians have grappled with the notion that Earth is not necessarily unique. The breath of life, bestowed upon the creatures of our world by either chance or design, may have equally well been placed upon beings of other domains. In an inestimably vast cosmos, the possibility that other inhabited places exist in space seems tantalizingly real.

Where are these new realms? The problem of locating alternative earths—worlds similar to our own—is a deeper and more challenging issue than the task of seeking out ordinary extrasolar planets of any type. It is one thing to seek out rocks in space that happen to have planetary proportions, it is another to quest for vital signs on faraway worlds.

Astronomers are particularly interested in seeking worlds that might harbor *complex* living organisms, perhaps even advanced, intelligent beings, rather than just simple, single-celled forms of existence. There are probably many bleak, hellish places in space, with horrendous environmental conditions, that could still support the presence of viruses and other microorganisms. But would you want to buy real estate on those planets? Sure, finding new locales where terrestrial amoeba might thrive would be significant, but we wouldn't want to call such worlds habitable.

A truly habitable world, according to most scientists' definitions, is one that supports a broad range of living creatures, from single-celled life forms, to beings as complex as terrestrial birds and mammals. It represents a place where evolution has taken—or at least could take—its full course, producing advanced, specialized beings. It need not be paradise, but it certainly should not be an abysmal wasteland.

None of the extrasolar planets found so far by globe-hunting astronomers seem to bear much resemblance to Earth. Therefore, if one makes the bold, far-reaching, and possibly erroneous assumption that terrestrial-like conditions are requisite for advanced life, then one might safely presume that none of the newly found worlds harbors complex living organisms. This supposition is bold and far reaching because it extrapolates what we know about our familiar domain—that it contains drinkable water, breathable air, and a comfortable range of climatic conditions—to potential new life-containing habitats in space. However, we know so little about what it means to be alive, aside from what it means to be alive on Earth, that this assumption may be entirely off the mark. Indeed, life-supporting worlds may exist that bear scarce resemblance to even the most barren parts of Earth. However, faced with no alternative but to rely on our familiar experiences of what life entails when we search for life in space, we are compelled to search for Earth-like planets in space.

By terrestrial standards, none of the recently discovered worlds seem potentially habitable. The pulsar planets found by Wolszczan, for instance, possess masses similar to Earth, but orbit a body radically dissimilar to the Sun. Like birds sitting under a timed fountain, periodically being sprayed with water, they are drenched with high frequency radiation every few milliseconds (without the protection of atmospheres).

One cannot conceive of life as we know it dwelling on these sorry orbs. While situated on their iron surfaces, there would be absolutely no shelter, no relief at all from the harsh conditions of pervasive radiation. There would be no warm sun smiling down, only the lethal stare of the pulsar eye overhead. Existence on such worlds would be worse than being a mouse trapped in a dentist's office beneath a continuously running x-ray machine.

It is similarly difficult to imagine life, by terrestrial criteria, existing on any of the worlds found through spectrography by Mayor, Queloz, Marcy, Butler, Cochran, and their colleagues. Although the giant extrasolar planets discovered by these astronomers orbit stars similar to the Sun, the erratic nature of their paths through space seems to preclude the possibility of favorable climactic conditions on their surfaces. Some of these bodies, such as 51 Pegasi B, hug their suns so closely, like moths pressed against a flame, that conditions are likely scorching. Others, such as those around 70 Virginis and 16 Cygni B, follow highly eccentric routes, weaving in and out of the temperate zones of their systems, and therefore cannot be expected to have stable, life-supporting climates.

Moreover, current surveys based on spectroscopy possess the capacity to find only giant worlds. They lack the precision necessary to detect planets of terrestrial dimensions. For this reason, all objects discovered by this method are monstrous in size. And, given what we know about conditions on Jupiter, Saturn, and other planetary behemoths in our own system, it is unlikely that mammoth extrasolar worlds, with their enormous gravitational forces, would prove fertile grounds for life.

Unlike spectroscopic searches, MACHO, PLANET, MOA, and similar gravitational microlensing projects have the capability of locating Earth-size planets. Yet, so far the evidence they have gleaned regarding potential planetary bodies has been inconclusive. They have seen a few striking glitches that may indicate the presence of massive bodies near stars. But in those cases, it has been unclear whether the phenomena seen represent proper planets or just diminutive stellar companions. Only time will tell if the microlensing projects will yield long-sought proof of the presence of Earth-size bodies around alien suns.

Where then, one might wonder, are those little green critters hiding? In what cozy corner of space are the colorful, flower-laden edens upon which many weary Earthlings dream of setting foot? Since all of the planets astronomers have found so far are believed to be barren—or devoid, at least, of advanced living organisms—then where might one hope to discover far-flung civilizations on inhabited worlds?

Let us not despair, however. Our search for habitable extraterrestrial realms has barely begun; one cannot expect victory so soon. Given human sagacity, there is every reason to believe that our quest will ultimately prove successful. We simply need to define what we are seeking

in the clearest possible terms, devote the necessary resources toward that aim, and hope for the best. Let us,
then, clarify what the quest for Earth-like worlds entails.

Vital Ingredients

Suppose you are the personnel director of a large
corporation. The well-respected vice president of the firm
has recently announced she is leaving for another company. She has been so successful in her present job that
she has been plucked by a rival business to become its
executive officer. Naturally, you and your colleagues are
sorry to see such a fine person jump ship. Although it will
no doubt be difficult, you have been assigned by your
boss to hire a replacement.

In order to focus the search before the interview
process begins, you decide to develop a list of necessary
characteristics that you would like all of the candidates to
have. Because you feel that the current vice president has
been ideal in every respect, you choose to base your
criteria on your perception of the best aspects of her
profile. You carefully write these down:

- Demanding—requires the best from every worker.
  However, not too intimidating
- Good sense of humor, but can be serious when
  necessary
- Energetic, but able to relax
- Highly moral, but not puritanical

Once you have compiled these guidelines, you post
an advertisement for the replacement position. In the ad,
you ask all applicants to write essays describing how they
would handle the job of vice president. Then you wait to
see what kinds of applications come in.

Several weeks later, you sift through the huge pile of responses to your ad. Each time you read an application, you take out your set of guidelines, and check off the ones that, based on the contents of his or her essay, the applicant seems to fulfill. In only the cases in which every criteria has been satisfied, do you bring in the candidates for an interview.

In any type of methodical search when the field of potential candidates is huge, the searcher typically applies a screening process to weed out poor choices. The quest for habitable worlds is no exception to this approach. There are trillions and trillions of stars in the universe. To examine each one carefully for planets, and each newly discovered planet for habitability, would take eons (if it were technologically possible). Therefore, as if they were conducting an interview to find alternative earths in the most efficient manner, astronomers must narrow their probe to only the most likely contenders.

The criteria used to narrow the pool of candidate planetary systems are based upon what we know about the limits of advanced life on Earth. These guidelines attempt to delimit in what sorts of habitats might complex living beings thrive. Experiment after experiment has shown that although simple organisms, such as viruses, might exist under a variety of extreme circumstances (low temperatures and pressures, for example), advanced beings require special environmental conditions to survive.

For human beings on Earth the range of livable climates is vast but not limitless. While Alaskan Inuits, living in the tundra, bear winter temperatures of $-40°F$, and Australian Aborigines, dwelling in the desert, tolerate summer heat of $+120°F$, obviously no one could survive in a land where temperatures regularly exceeded (or even came close to) $212°F$, the boiling point of water. Astronauts endure high accelerations during lift off, but they certainly could not survive forces 100 times that of gravity. And while deep sea divers are used to swimming under high pressure (literally, as well as figuratively), none would volunteer for a mission to the bottom of the Marianas Trench, the deepest point on Earth, 37,800 feet below sea level.

Considering what we know about terrestrial life's environmental limitations, and making the bold assumption (possibly erroneous, but necessary to focus the search) that alien organisms obey similar constraints, under what circumstances might we find habitable worlds in space? The conditions one might list pertain to what categories of stars the planets circle, the planets' orbital distances and types, their masses and spins, and their chemistries.

Seeking Suitable Suns

Let us begin this discussion by considering the types of stars that might support habitable planets. First of all, we expect that these stellar bodies must be on the Main Sequence (the zone of long-living stars). Stars on the Main Sequence burn hydrogen slowly and for a long period of time, generally providing a lengthy interval for planets to form around them, and a sufficient opportunity, at least timewise, for life to develop on one or more of those worlds. Main Sequence stars are also fairly stable in size, helping to insure that they don't enlarge too prematurely and gobble up a world where life has developed. It is not hard to find objects of this category; over 90 percent of stars in our galaxy are Main Sequence.

Furthermore, stars potentially supporting habitable planets must be of the right spectral class. The best possible stellar variety for that purpose is type G, of which the Sun is a typical member. Stars of this category, along with a fraction of those of types F and K, tend to shine steadily and brightly for billions of years, providing ample light and heat for potential life-bearing worlds in their vicinities. About 25 percent of the Milky Way's visible, luminous bodies fall into these preferred classes.

Hotter stars, those of types O, B, A, and some of type F, blaze too strongly. Within a span of 10–100 million years or so, they tend to radiate away all of the energy and then burn out. It is doubtful whether advanced living beings could develop near these kinds of stars in such short intervals of time. Cooler stars, on the other hand, such as those of type M (red dwarfs), put out too little energy to support life on nearby worlds. Thus, like Goldilocks in the classic tale by the Brothers Grimm, it is wise to eschew those that are too hot or too cold, and choose one (star type) that is just right.

There is another reason why type G stars seem to fit the bill. Unlike more erratic star types, they tend not to vary in their radiation output. Since sudden outbursts of energy would likely prove lethal to embryonic life, stellar stability would present a strong advantage. Advanced beings certainly could not emerge on a world in which fiery blasts from its sun regularly roasted its environment. For this reason, stable, Sun-like stars of class G would offer much greater hope of life.

Another important factor when assessing the likelihood of habitable worlds around a star is its approximate age. Stars that are significantly less than 4–5 billion years old—the age of the Sun—should be ruled out. Because evolution is such a painstakingly slow process, only planetary systems reasonably close to solar age would likely be mature enough to support advanced beings. Judging by Earth's example, 3–4 billion years seems to be the requisite minimum period for complex living organisms (and intelligent life in particular) to form.

Finally, stars under consideration for harboring life-supporting planets must possess the organic molecule-building heavy elements, such as carbon, nitrogen, oxygen, and iron that living tissue so critically requires. Not all stellar bodies possess these materials. Though younger stars generally do, some older stars do not contain any heavy substances, but are exclusively composed of light elements, such as hydrogen and helium. The distinction between these stellar compositions can be delineated by what astronomers call Populations I and II.

Comparing the two categories, Population II stars are much older, forged of the primordial gases that formed the early universe. In essence, they are hydrogen mills, converting simple hydrogen, the lightest element, into deuterium (heavy hydrogen) and helium. Through nuclear fusion, they burn these primitive fuels until their supply is exhausted, and they become extinguished. Once they die—typically in fiery explosions—their material is strewn throughout space, to be recycled by newer bodies.

Population I stars, in contrast, such as the Sun, are cosmic foragers. Like ravenous vultures, they feed on the substances discarded by Population II stars, digesting these elements in their churning stomachs to form more complex materials. Because they have scavenged the remains of older bodies, Population I stars possess eclectic compositions. Metals and other heavy elements, such as carbon, oxygen, and nitrogen, typically comprise 1–2 percent of the masses of such bodies.

No beings with which we are familiar are constructed solely out of hydrogen, helium, and similar light elements. Therefore, it is unlikely that habitable worlds could develop around Population II stars. Clearly, population I objects, comprising roughly one third of our galaxy, present far better candidates for systems that potentially harbor life.

Extant planet searches have taken these criteria strongly into consideration. Both the Geneva group, led by Michel Mayor, and the San Francisco group, led by Geoff Marcy, have focused their research efforts on stars that are very similar to the Sun in age, population grouping, stability, and spectral type. For example, 51 Pegasi, the first living star around which an alien planet was found (40 light-years away from Earth), is a veritable clone of the Sun. Planet-hunting teams will certainly continue to concentrate their efforts on probing Sun-like objects for nearby worlds.

How Close for Comfort?

Once astronomers have demarcated a suitable set of stars for scrutiny, they must grapple with the question of where near these selected objects might habitable planets likely be located. If the Solar System represents any guide, then surely complex life could not evolve on planets as close to their suns as sweltering Mercury, or as far away as icy Pluto. Rather, one might expect life to flourish on worlds situated at comfortable distances away from their mother stars and locked into stable orbits at these distances.

In the 1960s, researcher Stephen Dole, of the Rand Corporation think tank, proposed the term, "ecosphere," to describe the region around a star in which habitable worlds might reasonably exist. Presumably, the range of surface temperatures on bodies located within a star's ecosphere would include those that have proven suitable for life on Earth. Dole suggested that the Sun's ecosphere, for example, encompasses the shell located between .86 and 1.24 astronomical units from its center—corresponding to a variation in solar light intensity between 0.65 and 1.35 times Earth level. (One astronomical unit is defined as the average distance between Earth and the Sun.) Thus, Venus and Mars, according to Dole, lie, respectively, just in front of and just beyond this life-nurturing zone. Venus receives too much, and Mars too little, of the Sun's radiation to be habitable.

We now know that this picture is somewhat simplistic. Aside from the distance factor, one must also reckon with variations due to atmospheric conditions. In the familiar greenhouse effect, planetary atmospheres of particular compositions (of high carbon dioxide content, for instance) can serve as heat reservoirs, trapping radiation within their thermal blankets. The situation of Venus is a prime example of such an effect. Because of its thick atmosphere, full of carbon dioxide, solar heat that shines upon Venus is reflected back to its surface. Like swatted balls, ricocheting back and forth in the game of squash, the Venusian photons (light particles) are stuck within a closed environment, continuously bouncing between land and sky. Therefore its atmosphere, over time, has continued to grow hotter and hotter until it has been saturated with trapped energy. The sweltering conditions on Venus have precluded the development of discernible life forms.

If Earth had formed a few percent closer to the Sun, then many scientists believe that it too would have experienced a runaway greenhouse effect. Billions of years ago, its average surface temperature would have started on an upward path, until atmospheric conditions became too steamy for complex life.

On the other hand, if Earth were a few percent farther from the Sun, then perhaps the Ice Ages would have continued indefinitely. There is some chance that glaciation, the process of conversion of liquid water to frozen ice sheets, would have continued until all of Earth's oceans and rivers became locked up in ice.

It is indeed fortunate that Earth evolved in the Solar System's narrow zone between fire and frost. Because atmospheres can serve to amplify climactic disturbances,

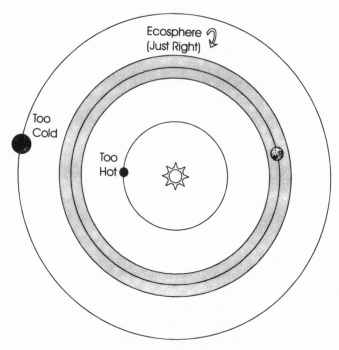

Figure 35. The ecosphere of a planetary system is the thin region in which average surface temperatures are ideal for life. Outside of this band, conditions are either too hot or too cold for Earth-like habitation.

small differences in radiant heat from central stars can potentially result in wide discrepancies in livability. In other words, take away some of Earth's heat, and it could rapidly freeze over; add some, and it could just as quickly boil over. Thus atmospheric conditions must be taken into account when stellar ecospheres are determined.

Another consideration pertaining to the location of planets around their suns is the magnitude of their orbital eccentricities (ovalness). If a planet follows an approx- imately circular route around its central star, it is far more likely to support stable climates than if its trajectory is greatly stretched out. A chicken barbecued on a (circularly) rotating skewer is more likely to be cooked evenly than one periodically brought closer to, and then farther away from, a hot flame. Thus, unless one seeks aliens that are crisp on one end and frozen on the other, in the ideal case for life, planetary orbits should have low eccentricities. That is one of the main reasons why the astronomical community holds out little hope for life on the highly eccentric worlds recently found around 70 Virginis and 16 Cygni B. Sun-circling (rather than sun-ovalling) planets would prove much better bets.

So far, we have only considered the case in which planets are in orbit around single, unaccompanied stars. But what of binary systems? Might habitable planets exist near a pair of stellar companions?

Imagine life on Earth if there were two suns in its vicinity. Suppose one of the stars was brighter and closer than the other. Each evening, a sense of tranquility would fill the air as the primary sun set in a blaze of colors. Evening songbirds would begin their nightly ritual as they perched on shadowy trees. The heavens would grow darker and darker, and all would become very still. But then, perhaps a short time later, a chorus of roosters would begin to crow. A false morning would begin, as the second sun broke through the darkness with its eerie dim glow. The heavenly dome would become somewhat lighter again as the twin rose in the sky. These patterns of rising and setting suns would continue, in and out of synchrony, as Earth turned each hemisphere toward one, and then the other, sometimes toward both and sometimes toward neither.

Isaac Asimov, in the classic science fiction story "Nightfall," pictures a world, called Lagash, that is situated in a system of six different suns. There is almost always at least one sun above so the people of Lagash are unfamiliar with the concept of night and are terrified of the darkness. Since it is always light out, they never see stars in the sky and are thus unaware of the greater universe. At the end of the tale there finally is an interval in which, due to their orbital configurations, all six suns cannot be seen. Darkness descends upon Lagash, sending the people into sheer panic. Riots erupt, cities are burned down, and the entire society is destroyed.

In reality, habitable worlds might emerge around multiple star systems only under a very special set of circumstances. Generally, for life to be feasible on a planet located in such a challenging environment, the stars in the system would have to be very far away from each other. The planet would have to orbit close to one of them and maintain a great distance from the other(s). One of the stars would be its primary source of light, heat, and gravitational influence; the other(s), being sufficiently far away, would have little effect. Only in this manner, with wide stellar separation, could the planet's path be simple and regular, and its climate be stable. Its turntable of seasons would revolve exclusively around the primary star, and not around its companion(s).

Otherwise, if a planet were engaged in a complex trajectory around several stars, then its meandering path through the system would likely render its climate too variable for life to evolve. Like a butterfly in a field of flowers, over time, the planet would flitter closer to one of the stellar objects, and then to another, its weaving path dictated by the laws of celestial mechanics. Consequently, its average surface temperature would vary drastically throughout its orbit. If, in its convoluted path, it moved closer to one of the hotter stars, temperatures would rise. If instead, it was drawn away toward a cooler star, or deeper into the space between the stars, then temperatures would plummet. Life would scarcely be able to adjust to being on such a fluttering planet.

There is a second, more remote possibility for stable orbits in a multiple system—particularly for a binary pair. Suppose a pair of stellar companions were extremely close together by astronomical standards: only a few million miles apart. Then a planet could conceivably orbit far enough away from the companions that it might follow a path reasonably close to circularity. Essentially, all environmental conditions would be the same for the world as if it were orbiting a single star, save that its occupants would always have two suns in the sky. Under that special set of circumstances, there would no reason to rule out the formation of advanced life on the binary-circling planet. Intelligent beings there, basking in the radiation of two suns, would think of their situation as normal—and would probably be surprised to hear about life on single-sun worlds.

Thus, only in the special cases of wide separation between the stellar companions and simple orbit around one of the stars, or extremely close situation and simple orbit around all of the stars could habitable worlds reasonably exist in a multiple star system. Otherwise, the chaotic nature of the planets' orbits would likely preclude the climactic stability needed for life to thrive.

The Right Slant

Two other significant factors that astronomers consider when they are surveying a planetary target are its slant and its spin. Only objects with equatorial inclinations (slants) and rotation rates (spins) similar to Earth are thought to be most suitable for life. If these physical parameters are too great or too small, relative to Earth's values, a world under consideration would have a reduced chance of habitability.

Like a mountain goat rounding a steep slope, the Earth is tilted as it moves around the Sun. Its equatorial inclination, the angle between its equator and its orbital plane, is a significant 23.5 degrees. The magnitude of this angle is the reason why, away from the equator, nights are so much longer in summer and shorter in winter.

Any variation in Earth's tilt angle would dramatically affect its climate. If Earth's inclination were a flat 0 degrees, then days and nights in every region would have the same duration of 12 hours each all year round. Seasonal differences would be nil, because every region at every time would receive essentially the same amount of daily sunlight.

On the other hand, if Earth's tilt were drastically increased—say, to 90 degrees—then the entire northern hemisphere would bask in perpetual sunlight for half of the year, and cower in pitch darkness for the other half. (The same would be true for the southern hemisphere as well, only in reverse order.) In that case, conditions would be simmering hot, beyond human experience, for six months out of the year and freezing cold for the other six months. It's hard to imagine living organisms flourishing in such a radically fluctuating environment.

With the extreme climatic shifts that a large equatorial inclination would produce, it is likely that most habitable worlds would have low tilt angles. It would therefore eliminate few viable prospects if one restricted one's search to inclinations of less than 80 degrees. (This is hardly a severe restriction; in our own Solar System, all planets except Uranus fall within these bounds.) Even at this upper angular limit, life would probably be restricted to a narrow belt around the planet's equator, encompassing less than one-tenth of its surface. For smaller tilts, the zone of potential habitation would be correspondingly bigger. Finally, for equatorial inclinations of 0 degrees, the entire planet would likely have an even climate and could theoretically support life, assuming that all other factors proved favorable.

Another consideration for habitability is the planet's rate of rotation about its own axis. Biologists have known for years that the Earth's 24 hour spin cycle sets the pace of existence for living beings on its surface. Each earthly creature exposed to the rising and setting of the Sun maintains an internal biological clock, known as the mechanism of circadian rhythms, that keeps similar tempo to the coming and going of day and night. These daily patterns include cycles of sleeping and wakefulness, appetite, and concentration. Even when an organism is removed from direct exposure to the outside world, these rhythms tend to continue at about the same rate.

Naturally, then, if Earth's rate of rotation were somewhat faster or slower, the pace of life on the planet would correspondingly be speedier or more drawn out. A day of 25 hours or 23 hours probably wouldn't have made much difference for the way life came to evolve. We would simply be chowing down our breakfasts, lunches, and dinners according to a different daily schedule.

However, suppose that the Earth rotated instead at a exceedingly slow pace—say, once every 500 hours. Its daytime and nighttime periods would be so long that temperatures would fluctuate greatly between those intervals. Days would be tremendously hot, and nights unbearably cold. Think of Alaskan winter nights, followed by Saharan summer days, only much worse. Perhaps life could not have evolved under those circumstances.

On the other hand, if the Earth were spinning much faster than it is now—say, once every hour—violent weather patterns could have rendered it just as barren as if it were orbiting too slowly. Because of its exceedingly rapid rate of rotation, and corresponding swift daily temperature variations, hurricanes and tornadoes and such extreme atmospheric phenomena would continuously blow across Earth's surface, serving to stifle nascent forms of life before they had a chance to develop.

It would be difficult to state, with justification, hard and fast rules about whether or not to reject planetary candidates on the basis of their anomalous spins. With admission that there is considerable room for error, researcher Stephen Dole suggests eliminating from consideration those that spin faster than once every three hours or slower than once every 96 hours (four days). These restrictions, however, would hardly be severe. In our Solar System the former criteria would exclude no planets from contention, and the latter, only Mercury, Venus, and Pluto. Until the relationship between the daily cycle of the Earth and the evolution of life on its surface is better understood, astronomers should refrain from applying stricter guidelines than these. In current searches, only the fantastically fast or extraordinarily slow spinning planets should be considered likely to be uninhabitable.

The final test in our physical exam for planets concerns their masses. Because of their crushing gravitational fields, objects as hefty as Jupiter or Saturn would be unlikely to foster life. High gravity on a world in which life was trying to develop would stifle the growth of organisms and hamper their movements. On massive planets, atmospheric pressures would be enormous as well, forcing fledgling entities trying to rear their heads back down to the ground. Moreover, if the Solar System provides any indication of what is in space, bulky planets are likely to be gas giants: planets composed mainly of simple materials such as hydrogen and helium. Thus they almost certainly wouldn't provide the necessary chemical elements for life.

Like too much mass, too little mass would also present a problem. An ultra-light planet—of the size of Mercury or Pluto, for instance—would similarly be unlikely to win a habitability contest. Such a puny world would be too weak to maintain all but an extremely thin atmosphere. And without a thick layer of air above it acting as a shield, most of its liquid water would either boil away or freeze on the ground (in the case that temperatures were cold enough). Ultimately, none of the main environmental conditions requisite for surface life (air, liquid water, etc.) would be present. (However, underground life could possibly exist.) We might therefore rule out such tiny worlds from our life-giving planet competition.

To summarize the astrophysical conditions that may have a bearing on the existence of advanced life on other worlds, the ideal candidates should:

- Orbit stable, long-living stars that are of spectral class (G) and population (I) similar to the Sun;
- Reside in systems that are at least 3–4 billion years old;
- Revolve around their suns at moderate distances— close, but not too close—with Earth's orbit as the model;
- Have near-circular paths, or at least ones that are not too stretched out;
- If in multiple star systems with widely separated stars, possess simple orbits around one of the stars, far away from the others;
- Have low-to-moderate equatorial inclinations;
- Have spins reasonably close to Earth's 24 hour cycle;
- Have masses similar to Earth; certainly not as light as Mercury or Pluto, or as heavy as Jupiter or Saturn.

The Elusive Blue Sky

Aside from its astrophysical characteristics, such as mass, spin, and orbital properties, there is another important way of determining whether or not a planet might have the capacity to support life: its chemical composition. In particular, if its atmosphere is abundant in molecular oxygen, ozone, methane, and other gases related to life-giving processes, then its prospects for habitability should be considered superior, all other factors being equal.

At first glance, the chances of testing the chemical makeup of a distant planet's atmosphere might seem slim. The closest known extrasolar worlds are dozens of light-years from Earth. Current technologies certainly would not permit us to launch probes successfully to locales such tremendous distances away. It would be millions of years, at current spaceship speeds, before they would arrive to perform measurements. Undoubtedly, such missions would be highly unfeasible. But without direct sampling of a body of air, one might wonder how its composition could be assessed.

Fortunately, in 1980, Tobias Owen of the State University of New York at Stony Brook suggested a powerful means by which the ingredients of extrasolar planetary atmospheres might be remotely determined. Owen, in an influential research paper, proposed that spectroscopy be used to search for life-sustaining substances, particularly oxygen, on worlds outside of the Solar System. Spectroscopy is the process by which incoming light is broken down into its component wavelengths. These wavelength patterns are compared to the signals of known chemical substances. Then, on the basis of this analysis, the composition of the light source can be resolved.

This process is analogous to that of hearing a familiar, but unidentified, song on the radio and trying to figure out its name and composer. If someone has a good ear for music and is trying to figure out the identity of a tune, he might listen to it carefully, jot down its musical notes (as he perceives them to be), and thumb through pages of sheet music to see if any known songs match up. A computer, with a data bank containing the notes from thousands of songs, could perform this task more quickly by comparing each song's musical score to that of the one heard on the radio, and seeing if any coincide. Similarly, chemical elements and compounds can be readily identified by their characteristic spectra. A detailed comparison to known spectral patterns can generally reveal the composition of a light source.

To perform such spectroscopic analysis, however, astronomers needed to be able to distinguish planetary light from the much stronger glow of their parent stars. Owen recognized that the technology required to pinpoint these precise signals from planets had not yet been developed to the proper extent. Undaunted by the constraints of the astronomical techniques of his time, he called for a concerted effort to distill and analyze planetary spectra.

Taking up Owen's rallying call, Bernard Burke of the Massachusetts Institute of Technology, as well as Roger Angel, A. Y. S. Cheng, and N. J. Woolf of the University of Arizona's Steward Observatory, published in 1986 detailed methods for sensing Earth-like planets by means of their spectra. Their procedures involve looking at the radiation of stars—and of possible worlds in their vicinity—by means of an optical technique called interferometry.

Interferometry is the science of analyzing the fringe patterns produced in interference: the phenomenon of light interacting with itself. Interference can be observed when an optical image is found to contain alternating zones of bright and dark fringes (similar to zebra stripes). These fringes are generated when various light waves, each with its own set of peaks and valleys, separate outward from a single source and then are brought together again (through focusing mirrors, for instance).

The brightness or darkness of the bands produced depends on whether or not the waves, when they recombine, are in or out of phase with each other. If the reunited waves are in phase—that is, peaks merge with peaks, and valleys with valleys—then bright bands are produced. This radiative addition process is called constructive interference. On the other hand, if the reunited waves are out of step with each other, and peaks from one set cancel out valleys from another, then dark bands are generated. This cancelling out of waves is known as destructive interference. By analyzing the configurations of bright and dark fringes created by a source, observers can use a set of mathematical relations to determine its diameter, spectral composition, and other optical properties.

Interference might best be understood by imagining two great ocean waves (tidal waves, perhaps) meeting each other from opposite directions and then colliding. If the crests of the waves happened to merge together at the

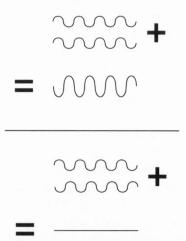

Figure 36. Depicted here is the process of interference. Two identical waves, added together in phase, result in a doubling of amplitude (maximum height). However, two identical waves, combined out of phase, cancel each other out.

same place and the same time, they would unite into a much bigger crest. This building up of waves would represent constructive interference (the equivalent of a bright light fringe). In contrast, if the crest of one wave happened to hit the trough of another, then the two would cancel each other out, and destructive interference would result (the equivalent of a dark fringe).

Burke, Angel, Cheng, and Woolf proposed using the interference of light from a suspected planetary system, as it is imaged by space-based telescopes, to detect and spectroscopically analyze Earth-like planets in the system. The placement of the telescopes in space would serve to eliminate the twinkle effect of atmospheric distortion. Also, in space, equipment can more effectively be kept extremely cold—freeing systems of the noisy (random signal producing) influence of background heat.

In Burke's model, a system of ten telescopic mirrors, spanning an area of 150 square feet, would act to collect incoming planetary light, recombine it, and allow it to interfere with itself to produce fringe patterns. In Angel, Cheng, and Woolf's variation of this method, a single large space telescope, with a mirror at least 33 feet in diameter, would be employed to produce a similar effect. In both cases, computers would analyze the spectral patterns of the incoming light, and compare them to those of familiar life-supporting substances, such as oxygen.

One of the main challenges of detecting Earth-like planets is being able to distinguish their signals from the powerful illumination of the stars they orbit. Angel and his colleagues addressed that issue in two ways. First of all, they suggested observing infrared radiation (heat emissions), rather than visible light. Because of their moderate temperature ranges (compared to stars)—between several hundred and several thousand degrees above absolute zero—planets emit far more infrared than visible radiation. Consequently, in the infrared spectrum, the typical intensity ratio between stellar and planetary light is only 10,000 to 1. This figure may seem high until it is compared to the characteristic ratio between stellar and planetary light in the visible part of the spectrum: a staggering 100,000,000 to 1. Thus, it is at least 10,000 times easier to observe a planet using infrared radiation, rather than visible illumination.

The second suggestion proposed by Angel's group was the notion of looking for the bright signals of a planet against the background of one of the dark fringes of its parent star's interference pattern. By adjusting their equipment to precise settings, they calculated they could place the image of an Earth-sized planet within the relatively dim zone of a stellar dark interference fringe. Since, at that point, the stellar background would be reduced in intensity by destructive interference, the planet could more readily be detected. In this manner, they predicted that it would be possible to record the spectra of distant extrasolar worlds.

Competing with Darwin

The application of infrared interferometry to planet hunting has been off to a slow start, outpaced by groups engaged in radial velocity measurements and gravitational microlensing. The reason that developments in this field have been so tortoise-like mainly has to do with funding. To do things right, infrared interferometry has to be carried out in space, free from atmospheric interference and of the noise of terrestrial background heat. In space, it is more efficient to keep telescopic mirrors cooled down to the low temperatures (less than 80 degrees above absolute zero, as Angel's team suggests) where the infrared radiation can most effectively be observed. But launching telescopes into space is an extremely expensive

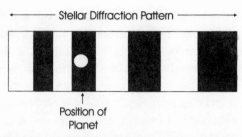

Figure 37. Theoretically, the dim glow of a planet might be seen against the backdrop of one of the dark bands of a stellar diffraction pattern. Diffraction, like interference, is an optical phenomenon generated by the merging of light waves.

endeavor, one that requires massive amounts of government support.

Currently, there are a number of worthy infrared planetary search projects vying with other experiments for competitive research grants. Perhaps the most prominent of these is the Darwin project, a proposal to the European Space Agency. Led by Alain Leger, of the Institute for Spatial Astrophysics in Paris, and including researchers from all over Europe, the Darwin project's aim is to search for Earth-like worlds around nearby stars, and then to scan the infrared radiation of any planets detected for the characteristic spectral patterns of water, carbon dioxide, and ozone. These substances were chosen because water is believed to be essential for life, and ozone and carbon dioxide are strongly associated with living processes.

Assuming that their project is funded, the Darwin group hopes in the next decade to launch an infrared interferometer in space, consisting of five 3-foot components, spaced more than 170 feet apart from each other. To reduce the distorting effects of interplanetary dust, they plan to propel their device out to a distance four times farther than Earth is from the Sun. Hoping that the telescope will be operable for at least four years, they expect to use the first year of its service to look for Earth-size extrasolar planets. The remaining years of the project would be devoted to spectroscopic analysis of the planets that they hope to discover. By the end of the mission, Leger and his colleagues hope to have found concrete proof of the existence of Earth-like extrasolar worlds with atmospheric chemistries that may support life.

Competing with Darwin, in a battle for the survival of the fittest international planet-hunting proposals, are a number of strong American efforts. Among these is the ExNPS (Exploration of Neighboring Planetary Systems) study, part of NASA's new Origins Program. The Origins Program has as one of its primary goals the resolution of the question: "Are there other worlds in the universe capable of supporting life?" Besides ExNPS, it includes a grab bag of other astronomical proposals, including the Space Interferometry Mission—designed to revolution-ize the field of astrometry (stellar wobble mapping) by conducting it in space.

ExNPS, the most recently proposed component of Origins, would be in many ways similar in scope and purpose to the Darwin project. It would also be extremely expensive, costing billions of dollars. Therefore, although ExNPS would be launched by NASA, and Darwin by the European Space Agency, I would wager that only one of the missions ultimately ends up getting full funding (and I hope that at least one does). Perhaps the best features of each project will be combined into a single international endeavor.

In its current formulation, the keystone of ExNPS is called the Planet Finder spacecraft. Consisting of four infrared telescopes linked together as an interferometer, it would be launched around 2005 into the relatively dust-free region of the Solar System beyond the orbit of Jupiter. Like the Darwin project, it would operate there for several years, first searching for Earth-like worlds, and then spectroscopically analyzing those found for evidence of carbon dioxide, ozone, and water vapor. NASA scientists hope that the importance of the information gathered will prove that space-based astronomical research is worth the cost.

Because space missions are so expensive, researcher Alan Stern, of the Southwest Research Institute in Boulder, Colorado, has recently suggested using ground-based interferometry, with the Keck I and II telescopes in Hawaii, to detect young Earth-size planets. Ordinarily, astronomers would expect that only space-based interferometers would have the power to locate objects of terrestrial proportions. However, Stern has shown, in a 1996 report, that recently formed planets would be much hotter than older worlds.

Young planetary systems are thought to contain far more asteroids and other debris that older ones. That is because, in newly formed systems, large quantities of the rocks and ices that occupy interplanetary space have yet to be captured by planets. Thus, for young systems, the chance of collision between asteroids and planets is quite high.

Continually bombarded with interplanetary debris, Stern estimates that an infant Earth-size world would glow 10,000 times hotter than a current one. Therefore, he argues that it would potentially be bright enough to see, even with a ground-based system. NASA is presently considering using the Keck telescopes for such a search.

Another American proposal to search for Earth-size planets is not based on interferometry at all, but on the related science of photometry. Photometry is the study of the intensity of light—in this case, that of stellar light. When a planet passes in front of a star, it casts a slight shadow, reducing the stellar intensity by a minute factor. For an Earth-size object, the reduction would constitute a 0.01 percent dip in the brightness of a star. Theoretically, this passage, called a planetary transit, could be detected by astronomers, revealing the presence of new worlds.

William Borucki, of the NASA Ames Research Center, has proposed continuously monitoring 8500 solar-like stars to look for Earth-size planets in their ecospheres (potentially habitable zones). His photometric design requires a 660-pound, 5-foot diameter telescope—equipped with thousands of light-sensitive CCD detectors—to be launched into space. He predicts that if other planetary systems are similar to our own, then 1 percent (a low, but significant value) will register transits with his method. According to his estimates, three to four years of monitoring would be needed to observe objects circling stars at the distance that Earth revolves around the Sun. Borucki's proposal, originally called FRESIP (Frequency of Earth-Size Planets), and recently dubbed the Kepler Mission, awaits funding from NASA.

Planetary Disk Pursuit

Arguably the current centerpiece of the global space observation program is the Hubble space telescope. One might wonder why such a powerful instrument that has produced spectacular pictures of Pluto, for instance, has yet to find direct evidence of a single extrasolar planet. Instead, other telescopes have been assigned to this worthy task. The truth is that the Hubble, though mighty, does not have the imaging power to observe, in a direct manner, distant worlds beyond the Solar System.

Yet, in January 1996, Dr. Christopher Burrows of the Space Telescope Science Institute in Baltimore, was pleased to announce the Hubble's first possible *indirect* sighting of an extrasolar planetary system. Using it to examine a dust disk that surrounds the star Beta Pictoris, Burrows discovered a curious bulge—a warping of the

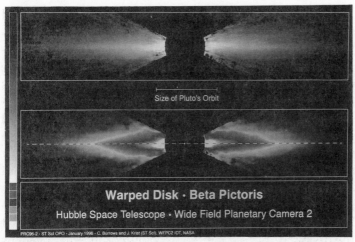

Figure 38. The warped disk of Beta Pictoris, as photographed by the Hubble space telescope's Wide Field Planetary Camera 2. Note the hole in the disk's center that represents the region where planets may be present. The size of Pluto's orbit is indicated on the photo as a basis of comparison; the disk's hole is somewhat smaller. (Courtesy of NASA)

disk that he believes was produced by the gravitational influence of at least one unseen planet.

Beta Pictoris is a rather remarkable object, located 50 light-years away in the constellation Pictor (Painter's Easel). It is a Main Sequence star, slightly hotter than the Sun. In 1983, NASA's Infrared Astronomy Satellite (IRAS), in a comprehensive infrared mapping of the Milky Way, revealed evidence that it was surrounded by a disk of gas and dust. Though, in the same survey, other stars were similarly found to have such disks, Beta Pictoris's was the only one that was later visually observed from Earth.

For the younger stars observed, theorists strongly suspect that these dusty veils constitute protoplanetary disks: grainy particle soups that will later coalesce together into larger and larger chunks, and eventually form planets. In the case of Beta Pictoris, an older star, they believe that a genuine planetary system has already emerged. The dusty disk, then, comprises the fine materials that were left over after planet formation.

Supporting the theory that Beta Pictoris has an existing solar system is the presence of a large empty ring—almost completely free of dust—that lies to the interior of the disk, and surrounds the star. The hole renders the disk more like a doughnut than a pancake. Researchers think that the lack of dust in the empty region indicates that planets have formed from the material once there.

The Hubble space telescope's contribution to the study of Beta Pictoris took place in January 1995. Using its Wide Field Planetary Camera 2, Burrows photographed the disk and its hole. Over the course of the year, he analyzed the image, looking for possible signs of deformation. Eventually, based on his investigation, he reached the conclusion that the disk was warped by a possible unseen planet or set of planets. By the start of 1996, he was certain enough about his results to make an announcement of his discovery at the San Antonio, Texas, meeting of the American Astronomical Society.

The warp's size allowed Burrows to make an estimate of the possible mass of the new planet. Based on the gravitational deformation that it caused, he calculated it to be between one-twentieth to 20 times the mass of Jupiter. He postulated that it could not lie too close to its sun, or else it would have been detected by radial velocity observation teams, such as Marcy and Butler's group. Therefore, he speculated that it occupies approximately the same orbit as Jupiter does around the Sun. According to his calculations, that would indicate that it possesses roughly the same mass as Jupiter as well. Though, because of its large mass, there is little hope that it would be a habitable object, the astronomical community awaits further confirmation of Burrows's likely planetary discovery.

Could Burrows's method potentially find an Earth-size world as well? Perhaps. With further refinement, as instrumentation becomes more precise, the minuscule warping of circumstellar (star surrounding) disks, induced by planets of terrestrial proportions, could possibly be detected. Astronomers hope that the Hubble space telescope, as well as ground-based instruments, will soon provide additional visual images of the dusty disks that are thought to surround at least one third of all stars in our galaxy. Then the quest will begin to sift through massive quantities of collected telescopic data and discover long-awaited proof of distant Earth-like bodies.

Chapter 8
Making Contact

We've only achieved the capacity for radio astronomy in the last few decades ... in a Galaxy where the average star is billions of years old. The chance of receiving a signal from a civilization exactly as advanced as we are should be minuscule. If they were a little behind us, they would lack the technological capacity to communicate with us at all. So the most likely signal would come from a civilization much more advanced.

CARL SAGAN, *Contact*

Ear to the Stars

Deep within the lush interior of Puerto Rico is a large, bowl-shaped valley, carved over the eons out of the rain-drenched soil by the powerful forces of natural erosion. Constructed on a network of steel cables just above the floor of the glen, taking advantage of its inverted hemispherical shape, is the largest antenna in the world, the 1000-foot diameter Arecibo Radio Observatory. One can-

Figure 39. The largest antenna in the world, the 1000-foot diameter Arecibo Radio Observatory. (Courtesy of SETI Institute)

not imagine a more exquisite site for a telescope. Under its giant dish, orchids and begonias bloom, while surrounding it lie acres of ferns, moss, and tropical grasses. A natural sink under the receiver drains excess water from tropical rainstorms, emptying through serpentine rivulets. Like the Golden Gate Bridge, the Grand Coulee Dam, Neuschwanstein Castle in Bavaria, and a handful of other spectacles around the world, it is a luscious combination of human-made wonder and natural splendor.

The folks at Arecibo are justifiably proud of their region's beauty. They are even prouder of the many "firsts" associated with their observatory. Every accolade bestowed upon Alex Wolszczan and his work reminds them that their site was home to the first discovery of extrasolar worlds, the pulsar planet findings confirmed in 1995. But their association with the quest for extraterrestrial realms in space dates back much further. The mighty Arecibo radio dish was used in 1974 to broadcast the first transmissions intended for alien ears, a series of messages (including profiles of our DNA and a map of the Solar System) sent to the M13 global cluster, 25,000 light-years away, to inform any intelligent entities that might reside there of our existence. Though this was the only time it was used to send messages, over recent decades it has often been employed to try to capture possible alien signals. In this manner, Arecibo has played a pivotal role in SETI, the Search for Extraterrestrial Intelligence. The connection between Arecibo and the quest for alien life will be cemented further when a powerful new procedure to scan the skies for extraterrestrial communications, called Project Phoenix, goes on line in 1998. The world community will then await another possible Arecibo first: the first conclusive radio evidence that intelligent beings exist outside of Earth.

In many ways, Arecibo is a symbol of humankind's interest in detecting, contacting, exploring, and perhaps even one day settling, habitable worlds circling alien suns. The research performed there embodies the two major ways scientists hope to establish whether or not there are Earth-like planets in the cosmos.

One way is the telescopic search for star-orbiting objects, with the hope of detecting ones that resemble our world as closely as possible. Detected at Arecibo, the planetary system found by Wolszczan scarcely looks like ours, yet its discovery lent a boost to the quest for more likely candidate earths.

The second way of searching for living worlds is to attempt to establish radio contact with the possible extra-terrestrials residing in the galaxy's habitable reaches. Radio communication, with its remarkable ability to span the inordinately vast gulfs of space, presents the ideal means for our race to attempt communication with others—and probably the same for them with us. For that reason, SETI has focused on scanning radio bands for possible alien signals, and, to a lesser extent, to relaying messages out into space. And Arecibo, as the biggest radio telescope in the world, has been a critical part of this endeavor to establish contact.

Ideally, the best approach to finding habitable extra-solar domains would be to combine these two methods into a single strategy. First, astronomers would continue to use assorted planet-hunting techniques—astrometry, spectroscopy, lensing, photometry, interferometry, etc.—to establish likely locales for advanced alien life forms. Then, once these favorable domains are pinpointed, SETI experts would conduct comprehensive radio scans of these sectors, hoping to detect messages sent by intelligent beings.

Project Phoenix, a seven year mission to detect evidence of extraterrestrial signals, will involve such a multi-part strategy. The planetary data gathered by researchers such as Cochran, Mayor, Marcy, and others, will form one component of the means by which the project planners will choose targets of their radio scan. If an Earth-like planet is soon discovered, it is a safe bet that the Arecibo dish (when it goes on line for Phoenix in 1998) will be directed at least part of the time toward the newly found body. The Phoenix team hopes that the strength of this multifaceted method will yield long-awaited evidence of intelligent life in space.

The Wizard of Ozma

Project Phoenix represents only the most recent in a long line of endeavors pertaining to the search for extra-terrestrial intelligence. Its design incorporates what we have learned in recent decades about the planets. Before we discuss this novel and important project, let us examine the history of SETI.

Though sporadic attempts were made as early as 1924 to detect radio signals from extraterrestrials—in that case, from "canal-building" Martians—SETI as an organized science is said to have begun in 1959. That year, "Searching for Interstellar Communications," was published by Giuseppe Cocconi and Philip Morrison in *Nature*. It calls for a methodical search for alien messages. Cocconi and Morrison advocated a sweeping investigation of radio signals from other civilizations. According to their estimates, radio waves constitute the most effective part of the electromagnetic spectrum for interstellar communications. Unlike other frequencies of radiation, radio signals can travel unhindered for trillions of miles.

A radio station on Earth, with enough power, could air a Beatles song to a receiver near Alpha Centauri. Granted, it would take four years after transmission before the song was received. If the song were sent in 1964, at the height of Beatlemania, it wouldn't be until 1968 that the music and lyrics of the Fab Four became known to our stellar neighbors. By the earliest time Paul McCartney or Ringo Starr could have received a fan message from them—in 1972—the Beatles had already broken up. At least the radio waves would have efficiently conveyed their music across space. (Perhaps their song "Love Me Do" is still on the charts somewhere in the galaxy.) Such is the ideal nature of radio for communications.

Cocconi and Morrison realized that scanning all possible radio frequencies for interstellar signals was simply not an option at that time. There would be far too many channels to investigate. Rather, they suggested that a search be centered on a frequency familiar to chemists and physicists: the 1420 megahertz (21 centimeter wavelength) line of atomic hydrogen. They assumed that because hydrogen is such a common element this frequency channel must be known to all intelligent races in space. Thus they supposed it would be the most likely avenue for communication.

They concluded their article with a rousing call for action that has served as inspiration throughout the years to the SETI program: "The probability of success is difficult to estimate, but if we never search, the chance of success is zero."[1]

Almost immediately, young radioastronomer Frank Drake, of the National Radio Astronomy Observatory in Green Bank, West Virginia, responded to this rallying call. By scraping together donated equipment and leftover funds, Drake managed to initiate a rudimentary search for alien signals with the Greek Bank 85-foot radio telescope. The fanciful nature of his mission—to establish contact with exotic new life forms—caused him to dub it "Project Ozma," after Queen Ozma from the Oz books.

Drake justified his search for extraterrestrials by means of a now famous mathematical formula. The "Drake Equation" is an algorithm designed to calculate

Figure 40. Frank Drake, president and founder of SETI. (Courtesy of SETI Institute)

the number of detectable civilizations in our galaxy: that is, from how many worlds might radio emissions be discerned. Drake estimated this quantity to be the product of the following terms:

- the number of new, suitable (Sun-like) stars formed in our galaxy each year;
- the fraction of those stars with planetary systems;
- the average number of Earth-like planets in each system that fall within their stars' habitable zones;
- the fraction of those planets where life arises;

- the fraction of life-sustaining planets in which intelligent life evolves;
- the fraction of those planets in which intelligent beings develop the technology needed to communicate over interstellar distances;
- the average lifetime of such technological civilizations.

Drake estimated the value of each of these terms and multiplied them to compute the chances of contacting extraterrestrial life. He calculated that there were approximately 10,000 technologically advanced civilizations in the galaxy—a quantity that he found reasonable enough to justify his search. In making such an estimate, he realized so little was known about habitable worlds in our galaxy that many of the equation's factors could not be well determined. Nevertheless, he advanced his equation as a means of launching discussion of the issue.

And that it has. Over the years, hundreds of researchers have debated the values of the terms in the Drake Equation. Some optimists, such as the late Carl Sagan of Cornell, have used it to purport that there are possibly hundreds of thousands of civilizations in our galaxy within reach of communication. Others have argued for lower estimates.

As astronomer James Sweitzer of the University of Chicago points out, except for the first term—the rate of formation of Sun-like stars—scientists have no idea as to the correct values of the remaining terms, or even the knowledge to estimate these properly. I would agree with this statement, with the exception that recent discoveries are bringing the second term—the number of planetary systems in space—well into focus. However, until astronomers discover Earth-like worlds, and adequately survey their properties, the value of the Drake Equation will remain indeterminate.

In April 1960, with high hopes for success, Drake

aimed the Green Bank telescope at two different nearby stars chosen for their Sun-like properties as well as their proximity. The first was Epsilon Eridani, a type K star, 10.7 light-years away. The second, Tau Ceti, a type G star, resides at 11.9 light-years away. Taking the advice of Cocconi and Morrison, he scanned the radio frequency hydrogen spectral lines of those stars. Alas, he found nothing but interstellar noise.

During the 1960s and 1970s there were a number of SETI searches, about 40 in total, modelled after Drake's. Most were carried out in either the United States or the Soviet Union where SETI had the most popularity. All were severely limited in funding and in telescopic use time. Therefore each project could only scan a few locations over a few frequencies (usually around the hydrogen line) for short intervals—hardly a comprehensive sky search.

With these limitations, there was no follow-up to interesting signals. Whenever an unusual pattern was found in some particular region of the sky, it was too late to re-examine that sector for more information. Data was generally examined months after the experiments were over. So, by the time any noteworthy signals were seen— possibly indicating messages—the search equipment had long been packed away, and the research staff had long been scattered across the country, engaged in other astronomical projects.

None of the SETI groups wanted to announce prematurely the presence of an alien signal, until that potential broadcast had been checked and rechecked by several different teams, they hoped. Since confirmations of signals were impossible, each group was forced to reach the same conclusion that nothing of interest had been found. Thus, by the end of the 1970s, the SETI program was left embarrassingly empty handed—and empty pocketed as well.

To make matters worse, in the 1970s a mammoth

UFO craze swept the West. Erich von Daniken's popular book, *Chariots of the Gods*, led much of the general public to believe that aliens had already landed on Earth, thousands of years ago. Practically every ancient wonder, from the Great Pyramids to the Inca city of Cuzco, was attributed to extraterrestrial architects.

Much of the scientific community, forced to spend valuable time debating believers in "ancient astronauts" from outer space, decided to rid itself of any connection to the search for extraterrestrial intelligence. Consequently, the academic esteem, funding, and telescope time awarded to SETI sunk to an all time low. As an indication of how poorly many regarded the project, in 1978 SETI won Senator William Proxmire's "Golden Fleece Award" for foolish and wasteful government spending.

Greeting the Aliens

One highlight for SETI during its otherwise dreary 1970s was the first terrestrial broadcast intended for alien audiences. The message sent through Arecibo in 1974, in the direction of the constellation Hercules, was a potpourri of earthly information aimed at stimulating extraterrestrial interest in our planet. It contained blueprints for human DNA (describing its double helix shape) as well as a map of the Solar System, a list of the elements important for life, and a depiction of the human form. There has been no response to the signal.

As SETI enthusiasts are eager to point out, the Arecibo broadcast was hardly the first communication from Earth that aliens might have received. Starting in 1939, a continued volley of television programs has been sent out into space. Any one of these broadcasts would be of sufficient intensity for aliens to have detected. Of course,

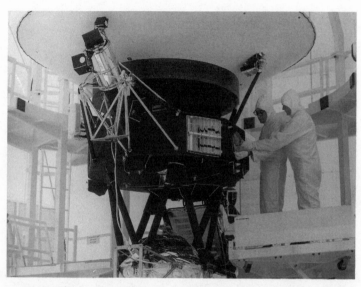

Figure 41. The "Sounds of Earth" record being mounted on the Voyager 2 spacecraft before it was launched toward the outer planets. The gold-plated copper record was encased in a gold-plated aluminum shield to protect it from the damaging affects of cosmic rays. Scientists hope the record will eventually be discovered by extraterrestrials. (Courtesy of NASA)

they would have needed to decipher and make sense of the signals. Being intelligent creatures, they may very well have developed the capability to appreciate 1970s situation comedies and the like. Perhaps the series "Mork and Mindy" (a comic television depiction of an alien) is now a hit somewhere in the direction of Betelguese.

Another means by which aliens might learn of Earth would be through the plaques, records, and other materials transported aboard four unmanned space probes: Pioneers 10 and 11, and Voyager 1 and 2. These spacecrafts, with primary missions to explore the outer Solar System, had the secondary purpose of carrying terrestrial information out into deep space. The items sent out included musical recordings, personal accounts of life on Earth, anatomical drawings, and other earthly souvenirs of potential interest to extraterrestrials.

Unfortunately, space probes edge across the galaxy at an elephantine pace compared to light-speed velocities of radio signals. It would likely take millions of years for any of the craft to reach extrasolar planets, let alone inhabited civilizations. By then, there is at least some chance the human race will no longer exist to receive the reply. For this reason, scientists pin little hope on obtaining responses to the Pioneer and Voyager messages.

Lonely Silence

By the late 1970s and early 1980s, the dearth of SETI results triggered off a debate as to why extraterrestrials, if they exist, have yet to make their presence known. One would think that if there were thousands of advanced civilizations in our galaxy, at least one of them would have contacted us by now. As the great physicist Enrico Fermi once retorted (much earlier) in a discussion of the abundance of extraterrestrial life, "Where are they?" This simple, cutting rejoinder has come to be known as the Fermi paradox.

Faced with the seeming failure of SETI, various researchers have postulated reasons as to why we haven't been communicated with. Their approaches fall into several distinct camps, spanning extremes of viewpoint:

- There are no extraterrestrials. Human beings are alone in the cosmos.
- There are extraterrestrials, but they have not reached us yet (or perhaps cannot reach us) because of the vast distances between them and us. We need to search harder for their signals.

- There are extraterrestrials who can reach us, but they are not inclined to do so. Perhaps they are trying to shield us from the shock of contact. Or maybe we are too boorish for them and not worth their time. (Earthlings are gauche, they chortle; who would want to talk with *them*?) If and when they chose to reveal themselves, they will.
- There are extraterrestrials who could have reached us. However, they destroyed themselves—or at least their technological base—through nuclear war or ecological disaster, before they discovered how to travel through interstellar space. Perhaps some advanced civilization will eventually clear these hurdles, learn how to develop interstellar space travel, and then contact us.
- There are extraterrestrials who can reach us; in fact, they already have (through UFOs, ancient astronauts who helped build the pyramids, etc.).

Most scientists fall into one of the middle camps, that is, neither thinking that extraterrestrials are already here, nor excluding their existence altogether. However there are some respected thinkers, though in the minority, who believe that aliens have periodically landed on Earth in UFOs. And there is at least one who is thoroughly convinced that human beings constitute the only form of intelligent life in the universe.

In 1981, Tulane mathematician Frank Tipler's paper, "Extraterrestrial Intelligent Beings Do Not Exist," caused a veritable uproar among the SETI community. His thesis that intelligence is unique to Earth is antithetical to SETI's whole reason for existence. SETI advocates upon learning of the article could scarcely believe that a reputable scientist, such as Tipler, would seek to eliminate all hope for interplanetary contact—the goal for which they had been striving.

Figure 42. Frank Tipler (b. 1947), professor at Tulane University. (Courtesy of Frank Tipler and Paul Daigrepont Studios)

Tipler is a strong advocate of the Anthropic Principle: the belief that the specific nature of the universe—its size, age, composition, etc.—can be justified by the fact that if it was otherwise, the human race wouldn't exist. In other words, the cosmos is the way it is to provide a suitable environment for the human race. If the early universe was slightly different, then the conditions that led to the development of advanced beings on Earth wouldn't have arisen, and we wouldn't be here to speculate about our origins.

Implicit in this argument is that the human race, as an intelligent species, is extraordinarily special, needing highly specific prerequisites for it to have arisen. According to the Anthropic Principle, the way the universe has evolved has been ideal for intelligent life on Earth; chances are it would not be so beneficial for different entities on other worlds. Thus, the cosmos is like a outfit tailor-made for an individual with unique characteristics—say, 8 feet tall, but only 100 pounds, with immensely broad shoulders, but with a neck only 3 inches in diameter. Such a suit would probably not fit anyone else, except the person for whom it is designed. Similarly, the argument goes, it would be unlikely that other life forms in space would fit our universe which is, after all, perfect for humankind.

Tipler takes the implications of the Anthropic Principle one step further. Not only does he argue that intelligent life is unlikely to exist on other worlds, but also he purports that it certainly does not exist on other worlds. Much to the chagrin of enthusiasts such as Drake, he excludes the notion that SETI will ever prove successful.

Tipler draws his strong conclusion from the simple fact that aliens aren't here already. Any intelligent race on a distant planet, he argues, would inevitably develop the capacity for space travel. Soon, it would send out probes and attempt to colonize neighboring worlds. Eventually, it would have the capability of exploring the entire galaxy. Finally, the wave of exploration would reach Earth. Therefore, since extraterrestrials haven't reached us, they must not exist.

The means by which Tipler proposes that extraterrestrials would settle the galaxy is by self-replicating robot survey ships. He borrows an idea from the renowned mathematician John von Neumann that eventually robots will be programmed to fashion materials from their environments to build likenesses of themselves. In this manner, craft bearing "self-reproducing automata" would be able to move farther and farther outward into space from their home planet, manufacturing more and more copies of their kind. Eventually, like ivy planted on a moist wall and left to grow over time, the galaxy would be full of these automata.

In his article, Tipler estimates how long it would take for the entire galaxy to be explored by this method. Assuming an interstellar flight would take approximately 100,000 years to complete, and that it would take robots about 1000 years to establish camp on a new world, reproduce themselves (by gathering materials from that planet and then using them to fashion robot replicas, in the manner described by von Neumann), and then prepare for a new journey, he calculates that it would take a civilization 300 million years to cover the Milky Way. Tipler further estimates, based upon Earth's history, that it would take 6 billion years from a planet's formation until the time that its technologically advanced life reached the evolutionary stage of being capable of performing this feat. Thus, adding 300 million to 6 billion, he arrives at the figure 6.3 billion for the number of years that an intelligent-life-bearing planet would exist before its inhabitants have spanned the galaxy, and thereby reached Earth.

Over half of the stars in the Milky Way are at least 6.3 billion years old. So, if even a small fraction of these have planets harboring technologically advanced beings, then Earth would have long ago been contacted. Since, as far as we know, it has never been approached, then Tipler concludes that extraterrestrials do not exist.

A number of scientists have issued rebuttals to Tipler's conjecture. The small group that believes extraterrestrial spacecraft have already arrived (in the form of UFOs or otherwise) might agree with Tipler's basic analysis, but beg to differ with his conclusion. They would say that intelligent beings have indeed developed the capacity to travel through the galaxy and have made their presence known on Earth.

The majority of SETI supporters who believe that extraterrestrials are out *there*, not here, have taken a different tact. Vocal advocates of the existence of intelligent life in space, such as Frank Drake, the late Carl Sagan, and Michael Papagiannis (of Boston University) have postulated a number of reasons why alien beings have yet to reach Earth.

Drake, who is used to juggling tight budgets for SETI, speculates that fiscal conservatism may prevent aliens from wishing to explore the galaxy. He ponders reasons why interstellar travel may be economically impractical for extraterrestrials (and they may be choosing to send messages instead). They may consider long distance space flights to be costly and dangerous. Perhaps the bulk of them would prefer the safety, comfort, and low expenditure of enjoying life in their own planetary system instead. As Drake has remarked:

> The argument I favor is simply that it makes better sense to colonize in your own system than to endure the costs and hazards of going to other stars. There may indeed be enormous numbers of civilizations of great technical prowess that don't bother to come to Earth in person. In other words, I assume that an intelligent civilization will colonize space *only* if it gets a good bang for its buck, that is, only if the quality of life for its expenditure is equivalent to the lifestyle it would get for the same amount of resources, energy or money in its own system.[2]

Perhaps it is Drake's familiarity with congressional budgetary hearings that leads him to the conclusion that extraterrestrials would be disinclined to allocate funds and personnel for risky endeavors such as space colonization. Along these lines, one might picture on a distant planet called Flxxz the alien equivalent of Senator William Proxmire awarding the "Golden Flxxz Award" to an interstellar space exploration program, and then ordering a funding cutoff. "We should conserve our credit units and balance the Flxxzian budget instead," the extraterrestrial legislator might argue.

Sagan, who was more interested in the politics of war and peace than in the intricacies of economic planning, presented another solution to the problem of why no alien beings have been seen. Speculating that extraterrestrial civilizations might be grouped into warmongers and pacifists, he saw reasons why each might fail to reach Earth.

Consider first the warmongers. Gradually they have built up advanced technologies, only to devote the power of their achievements to warfare. Foolishly, they have decimated their ecologies, engaged in an endless pursuit for military victory. Yes, they have thought of exploring the stars and exploiting the riches of space. But before they have traveled very far, they have destroyed themselves in senseless global battles. Instead of establishing themselves as galactic leaders, they now suffer the radioactive nightmare of nuclear winter or else the devastating effects of chemical and/or biological warfare. Their overwhelming ambition has thereby led to their mass suicide. That is why they have never stepped foot on our planet.

Other races, Sagan imagined, might arrest their violence before making such a fatal mistake. These pacifistic cultures would dedicate their resources to healing, not hurting. They would spend their time and energy for medical, rather than military, advances. Maybe as a result of their efforts, they would attain extremely long life spans, or even immortality.

Such pacifistic, long-living beings would scarcely wish to conquer space and colonize the galaxy. For one thing, such ventures would smack of imperialism, a tendency that they have tried to avoid in their desire for peace and harmony. Also, space travel would be physically risky. Why chance the perils of an interstellar voyage, and possibly cut short one's long life, when home is so healthy, harmonious, and cozy? This manner of thinking might influence them to eschew such journeys and remain on familiar turf. Thus, they would never be seen on Earth (but may, instead, decide to send messages).

Papagiannis, one of the founders of the field of bioastronomy, has proposed a third possible rebuttal to Tipler's thesis. Perhaps extraterrestrials do have the ability, and even the will, to explore Earth. However, in the interest of not harming our culture by revealing themselves prematurely, they are holding back. They are refraining from contact until the human race is mature enough to handle the experience. Once we would no longer be shocked by an alien encounter, Papagiannis conjectures, then they will appear in droves to greet us.

Who knows? Any, or none, of these answers could be correct. It seems that the best solution to this debate is to follow the advice of Philip Morrison of MIT, who, with Cocconi, wrote the seminal paper that helped launch the whole enterprise. Morrison implores researchers to refrain from baseless speculation, and just get on with the search for extraterrestrial life. We have the technology to look for other-worldly signals, so we should exploit it to the fullest. Only when we have discovered proof of the existence or nonexistence of intelligent entities in other systems should we be satisfied. Thus, in Morrison's view, the quest must continue until we know the definitive answer to the question "Are we alone?"

Phoenix Rising

In recent decades, the SETI program has slowly made a comeback, albeit with some ups and downs. In the early 1980s, to soothe the sting of the 1978 "Golden Fleece Award," a number of well-respected scientists associated with the search for life in space launched a publicity campaign aimed at promoting their mission. They needed to act fast and hard; Senator Proxmire had launched a bill in the U.S. Congress aimed at prohibiting spending for SETI. Proxmire felt SETI was a waste of time and money because, in his opinion, it was obvious that life did not exist beyond Earth.

Sagan, president of a newly formed advocacy group for space exploration called the Planetary Society, took the lead in combatting the attack on SETI. He met with Senator Proxmire, spoke with him in depth about the implications of the Drake equation, and convinced him that the search for life in space was an enterprise worthy enough for at least some federal funding. Proxmire still believed that the human race is alone, but he was brought to see enough merit to the other side to justify allowing government supported research.

About the same time, Sagan circulated a petition in endorsement of SETI, which was published in the October 1982 issue of *Science*. Signed by 71 prominent international figures, including seven Nobel Prize winners, it called for a coordinated search program to test for the existence of extraterrestrial intelligence. Signatories included Linus Pauling, Stephen Hawking, Stephen Jay Gould, and Lewis Thomas, among others. The impact of the petition, coupled with the effect of Sagan's meeting

with Proxmire, earned SETI a Congressional appropria-
tion of $1.6 million for fiscal year 1983, the start of a five
year research program.

Throughout the 1980s and early 1990s, government
funds flowed into SETI like thick maple syrup, slowly but
sweetly. In 1984, the SETI Institute was founded, headed
by Frank Drake, to coordinate search efforts. A nonprofit
organization located in California, it worked with NASA
to make the best use of targeted research grants. Within
a decade, $58 million was spent by the United States on
the development of new equipment and methods for the
mission to detect other-worldly signals.

The culmination of these collaborative efforts was
the October 1992 launching of NASA's High Resolution
Microwave Survey (HRMS) project, timed to coincide
with the 500th anniversary of Columbus's landing in
America. It was planned to be a 10 year sky survey of
incoming radio waves over an unprecedented wide range
of frequencies and stellar targets. Drake, Sagan, and their
colleagues were delighted with the prospects of such a
concerted effort to seek extraterrestrial communications.

Then disaster struck SETI once again. Within a year
after the initiation of the HRMS, at a time when its team
members were still performing preliminary tests, Congress cut off all government funding. Those involved in
the project were crushed; a promising enterprise was
ending before it truly had the chance to begin.

Fortunately, by that time, SETI had attracted a number of loyal supporters who were willing to make strong
financial contributions to the project. The HRMS mission
was relaunched as Project Phoenix, a decade-long pri-
vately funded effort coordinated by the SETI Institute to
discover radio evidence of extraterrestrial intelligence.

Project Phoenix is directed by astrophysicist Jill Tarter. Tarter, who obtained her Ph.D. from the University of
California at Berkeley in 1975, is a stalwart searcher for

Figure 43. Jill Tarter, director of Project Phoenix. (Courtesy of SETI Institute)

radio signals from space. For her steady devotion to such a taxing cause, she has won two NASA Public Service Awards and a Lifetime Achievement Award from Women in Aerospace. She is also the Principle Investigator of the SETI Institute.

From the start, the scope of Project Phoenix has been well defined. Unlike many previous SETI runs, Phoenix does not concentrate its efforts on and near the 1420 megahertz hydrogen line (a region of the spectrum known as the "water hole" for its proximity to one of the spectral bands of water). Rather, its intended range of frequencies will be from 1.2 to 3.0 gigahertz—a portion of the spectrum known for its relative quiet. Signals sent over these bands would not have to fight with much background noise from natural astrophysical phenomena to make themselves heard.

Three distinct groups of stars will be surveyed by the team. The first category, called the Nearest 100 Sample, includes the 100 stars closest to the Sun, regardless of stellar composition. All of these stars are within a distance of 25 light stars. Familiar members of this group include Alpha Centauri (a type G star), Proxima Centauri (a red dwarf), and Sirius (a type A star, much larger and brighter than the Sun). Because many of the targets are very different from the Sun—and a large percentage are in multiple star systems (25 binary systems, 7 triples, and 1 quintuple), Phoenix researchers do not expect the first group to yield the best results. However, if signals are discovered within this range, they would have emanated from systems close enough to Earth that exploration would be most conceivable.

The second category, known as the Best and Brightest Sample, includes 140 stars within 65 light-years that are very much like the Sun in color, temperature, and size. Those that are too young or too old, or have proximate stellar companions, are ruled out. Close Sun-like stellar bodies, known to have planetary companions, are included on this list. This is the group for which, because of its relative proximity, Project Phoenix holds the most promise. Therefore, it will be observed for the longest amount of time.

The largest segment, called the G Dwarf Sample, comprises stars that have the same stellar type as the Sun and are less than 200 light-years away. Selection for this third category is stricter than for the first, but not as rigorous as for the second. Approximately 700 stars are included in this group of solar-size, type G objects. Because they are farther away from us than are the other targets, if any of the systems from the third category send messages, it will take many decades or even centuries for them to receive our response.

The opening phase of the Phoenix search took place during the first half of 1995 at the Parkes Observatory in New South Wales, Australia. Approximately 200 Sun-like stars, visible from the Southern Hemisphere, were scrutinized using the 210-foot diameter radio telescope. After a five month survey, the Phoenix observing system was brought back to SETI headquarters—first on a flatbed truck to Sydney, and then aboard a cargo ship to California—where the data taken was analyzed.

In September 1996, the group's equipment was moved to Green Bank, West Virginia, for a run at the National Radio Astronomy Observatory. There, they share the 140-foot telescope with a number of other radio astronomy projects. Consequently, their observations are limited to short intervals, ranging from a few hours to about a week at a time.

The last stage of Project Phoenix will be conducted at Arecibo Observatory. The 1000-foot telescope there is currently being upgraded. Once these improvements are complete, the instrument will offer radio astronomers greater sensitivity and a broader range of wavelengths. Phoenix researchers have already reserved large blocks of observing time, lasting at least until 2000. Because of the incredible scanning power of Arecibo, they expect that the final stage of their current project will prove decisive.

Preparing for Contact

An incalculable loss for SETI has been the recent death of one of its strongest supporters, Carl Sagan. Before he died in late 1996, Sagan expressed the hope that he would live to see proof of intelligent life on other planets. He even wrote a novel, *Contact*, imagining the first encounter between humans and an alien race. His enthusiasm for the search for extraterrestrial intelligence will be sorely missed.

In his lifetime, Sagan was occasionally chided by "serious" astronomers for presenting popular accounts of the field. What these critics failed to recognize was the long and important history of leaders in astronomy— from Arthur Eddington and George Gamow to Edward Harrison and George Smoot—issuing general audience presentations of their work. Only by making knowledge of important astronomical achievements accessible to the public will the field receive the support it needs to flourish. Consequently, Sagan's critical contribution to the promotion of space science should not be undervalued.

Sagan maintained that the human race must always stand ready for interstellar contact. Any day, he argued, we could find ourselves no longer alone in the cosmos. The billions and billions of bits of information transmitted over television broadcasts could readily be picked up by extraterrestrial receivers. They could then, in turn, send a signal to us, which could conceivably be collected and

Figure 44. Carl Sagan (1934–1996), Cornell University astronomy professor and founder of the Planetary Society. (Courtesy of Cornell University)

interpreted by SETI. Finally, in translating that message, we would suddenly realize that there are others in space. We must perpetually be alert for that pivotal moment.

To increase their readiness in responding to a likely extraterrestrial signal, SETI experts have prepared a set of guidelines for what they would do if contact seemed imminent. These regulations would help prepare the world for what would likely be the greatest shock in history.

First of all, after receiving an interesting pattern, they would attempt to decode it. SETI researchers believe that this wouldn't be difficult. In order to engage in radio communications, extraterrestrial civilizations would need to know basic science and mathematics. Presumably the coding systems used for messages would involve universal scientific and mathematical principles, and could thereby be translated.

Second, other radio observatories would be alerted as to the discovery. Confirmation of the signal pattern would be sought. Only if other groups witnessed the same configuration would further action be taken.

Next, if the extraterrestrial message is confirmed, a global announcement would be made. The SETI team does not believe in secrecy. They feel trying to cover up scientific information would only lead to suspicion and paranoia. All attempts should be made to allay public anxieties for these would only lead to global panic.

Ultimately, SETI experts believe that the best policy is to prepare society in advance for the possibility of an extraterrestrial encounter. At the same time, it is important to let the public know that evidence of alien intelligence has yet to be found. That way, if contact does occur, the general attitude will be neither overly anxious nor terribly lackadaisical. Consequently, society will be in the best position to contemplate possible responses to the message.

Reactions to a friendly communication from another planet would probably include sending a return signal. Astronomers would likely begin a heightened study of the region of space from which the message was sent. Finally, once the proper technology is developed, a space voyage to the extraterrestrial civilization might be attempted. And then our time as a solitary race would draw to a swift close.

Conclusion
PUSHING
THE LIGHT BARRIER

> *We, this people on a small and lonely planet*
> *Traveling ... past aloof stars,*
> *across the way of indifferent suns ...*

> MAYA ANGELOU, *Poem for the 50th Anniversary*
> *of the United Nations*

Like an infant falcon we sit perched on the lonely rock of Earth, our new-formed wings poised, waiting for the inevitable instant of freedom. Eyeing the incredibly vast expanse around us, a universe of virtually unlimited possibilities, we pine for other roosts, but see none in the haze. We cock our heads as we listen for sounds from afar, but all we hear, day after day, is silence.

One day, the right moment arrives. Now our eyes and ears have sharpened with maturity and are beginning to sense novel sights and sounds. Maybe it is the faint, distant signals, freshly permeating the air, beckoning us like sirens. Could they be vague messages from creatures like us? Or perhaps it the specter of far off hills, barely visible in the fog, that draws us closer. Might potential companions live there, eager to learn of our presence? Whatever has changed, whatever new has arrived, we know that our time has come.

We flex our wings and begin to press them together more and more quickly. We take a deep breath, then, without hesitation, lift off high into the air. Soon we have left our rock of birth behind and are soaring freely through space. We slip beyond the nest of the familiar, flying fast along the path of adventure—onward, onward, onward, toward the wonderful great unknown.

Ever since early humans watched birds glide through the heavens, it has been a long standing dream of our kind to fly. And at least since Galileo showed that it was possible to stand foot on other planets, we have fantasized about spaceflight. With the discovery of new worlds around distant suns, and the expansion of the search for intelligent life in these and other systems, our hunger to leave Earth's harbor and set sail for exotic places in space has grown even stronger.

Recent findings of Jupiter-size objects in space and strange planetary bodies around pulsars have represented late 20th century astronomical marvels. In coming decades we are likely to witness the first sighting of an Earth-size planet circling a Sun-like star, the first indications of atmospheric oxygen and water vapor on such worlds, the first signs of life in space, and perhaps even the first evidence of extraterrestrial intelligence. Information gathered from infrared, radio, and optical telescopes will almost certainly enlarge our understanding of habitable and inhabited domains in our galaxy.

Assuming funds are available and the public is eager, in centuries to come we will begin to contemplate interstellar voyages. Rocket technologies will continue to improve, and spaceship speeds will likely grow faster and faster. Finally, journeys to the nearest stars, such as Alpha Centauri, will be seriously considered.

A significant obstacle, though, to interstellar travel is

the severe limitation of the speed of light. As Einstein pointed out in his theory of special relativity, no object may accelerate from ordinary velocities to the speed of light (186,000 miles per second) or faster. Only signals already traveling at light speed (radio waves, for example) may continue to move at that velocity. The speed of light thus presents a natural upper limit to the efficiency of space voyages.

No space mission, no matter how powerful its engines, could reach the planet around 51 Pegasi in less than 40 years (Earth time). A round trip to and from that world, including a 5 year stopover, would represent a longer period than the average human lifetime. If astronauts somehow left for such a journey at the time Woodrow Wilson was president, they still would not be back.

Fortunately, another aspect of special relativity makes prospects for interstellar journeys a little brighter. Einstein proved that clocks on spaceships traveling close to the speed of light would tick at slower rates compared to those stationed on Earth. Thus, astronauts journeying through space at high velocities would age more slowly, throughout the duration of their voyages, than their friends and families back home. During a round-trip journey to 51 Pegasi, traveling there and back at 99.99% of the speed of light (with a 5 year stopover on the planet), an astronaut would age only a little more than 6 years, while his terrestrial companions would experience a full 85 years of life. This effect, called *relativistic time dilation*, makes interstellar journeys more feasible, by shortening the periods of time experienced by extremely fast moving astronauts. The technologies needed to achieve such speeds, however, are likely many, many years away.

A similar effect (of the shortening of the apparent duration of voyages), though at lower speeds, might conceivably be carried out someday through the science of cryonics. Cryonics, in this context, is the deep freezing of human bodies, with the intention of reviving them to full function at later dates. At very cold temperatures, human tissue does not deteriorate. Therefore, some scientists believe that "sleeping" (cryonically preserved) astronauts might potentially be restored to waking states after years, or even centuries, of stasis. They could thus avoid aging and boredom during long space voyages. A computer system aboard spaceships could be programmed to freeze, monitor, and later resuscitate astronauts as planetary targets grew near. Like ultra-high-speed space travel, the technology to perform such a feat seems ages away.

Another conceivable means of reducing tediously long space journeys would involve the warping of space. According to general relativity, mass concentrated in a particular region of space causes it to bulge as if it were rubber. The heavier the material in a given sector, the greater its distortion. Some theoretical models suggest that this curving process could be used to create a traversable wormhole (a kind of space tunnel) between one part of the universe and another, through which astronauts might embark on shortened journeys to extrasolar planets. Theorists who have advanced this view, such as Kip Thorne of Cal Tech, are quick to point out the extremely hypothetical nature of general relativistic shortcuts through space. Therefore, at this point, wormhole technologies should certainly not be considered on par with more conventional approaches.

Finally, to circumvent the limitations of interstellar space travel, Frank Drake suggests that information, not matter, should be exchanged. To save time, money, and lives, we should conduct our business with extraterrestrials via radio communications, rather than through lengthy, dangerous voyages. Our mode of greeting our remote new friends, he argues, should be "phone calls" rather than handshakes, that is, virtual, instead of real.

Drake undeniably has a point. Telephone, telefax, and internet technologies have shown us that being there is hardly essential for business and leisure. However, I feel that the human race pines for actually setting foot on other worlds, if at all possible, rather than just experiencing them from a distance. Radio contact will probably only be a prelude to genuine space voyages.

Given the rigors of interstellar space travel, if we do decide to attempt it (when such technology becomes available), we will surely choose our targets carefully. The worlds we'll explore presumably will be selected for their likelihood of habitability, determined by meticulous telescopic observation—or through radio communication, if contact has already been established. For safety's sake, robot probes will probably be sent out to planets of interest long before manned missions are dispatched. Finally, the treacherous journeys to novel extrasolar worlds would be carried out with the greatest possible care.

Not all planets of interest for exploration would have ideal conditions for human habitation. A certain percentage of less-than-habitable worlds, with some conditions favorable and others detrimental to human life, might present good candidates for "terraforming." *Terraforming* is the hypothetical process of transforming a planet so that it more greatly resembles Earth, mainly in terms of atmospheric conditions and climate. Terraforming expert Martyn Fogg, a dentist by training and a well-respected astronomer by interest, has drawn up detailed blueprints for such metamorphoses. He speculates that a combination of physical and chemical methods, such as placing shades into orbit to block an especially harsh sun, or releasing oxygen producing substances into a carbon dioxide-based atmosphere, could turn a hostile terrain into a livable environment.

Fogg has published detailed scenarios for the terraforming of Mars and Venus, which would involve painstaking efforts to transform their environments into habitable realms. He believes, however, that the terraforming of a young extrasolar planet would be much easier. As he related to me:

It seems to me that whilst Mars and Venus are, conceptually, very difficult to terraform at the present time, if we had had access to them during the first 1–2 Gyr (billion years) of their evolution, the task would have been much easier. It may be that examples of such "biocompatible," but not habitable, worlds may be quite common about other stars.[1]

Eventually, if interstellar travel is perfected, our terrestrial civilization may expand throughout the galaxy, encompassing newly colonized worlds—some already habitable, others rendered livable (perhaps through terraforming). Our planet may gradually lose its place as the primary dwelling place of the human race. In that case, future historians may well describe the recent discoveries of extrasolar systems by Wolezczan, Mayor, and others, as the beginning of a new era of spatial exploration and settlement.

Someday Earth will no longer be habitable. Billions of years from now, the Sun will eventually die. If the human race has the prescience to claim other worlds for its own, it might very well survive the inevitable extinction of its mother planet. That is the supreme hope that lies behind humankind's quest for worlds outside of the Solar System.

Like the falcon, we are creatures of freedom. We cannot forever remain confined to our earthly nest. With our keen telescopic vision, sharpened as we scientifically develop, we regularly eye future targets of flight—newly discovered extrasolar planets in space. When our manufactured wings grow strong enough, toward them we will delightfully soar.

REFERENCES

Preface

1. Arthur C. Clarke, *Childhood's End* (New York: Ballantine Books, 1953), p. 75.

Introduction

1. Frank D. Drake and Dava Sobel, *Is Anyone Out There? The Scientific Search for Extraterrestrial Intelligence* (New York: Delacorte Press, 1992), p. 1.
2. Frank D. Drake and Dava Sobel, Ibid.
3. Alexander Wolszczan, personal communication, June 10, 1995.
4. William J. Broad, "Scientists Widen the Hunt for Alien Life," *New York Times*, May 6, 1997, p. C1.

Chapter 1

1. Bernard de Fontanelle. Quoted by John Noble Wilford in the *New York Times*, February 25, 1996, p. F1.
2. William Herschel, "On the Nature and Construction of the Sun and Fixed Stars," *The Scientific Papers of Sir William Herschel* (London: The Royal Society and the Royal Astronomical Society, 1912), Vol. I, p. 156. Reported in William Graves Hoyt, *Lowell and Mars* (Tucson: The University of Arizona Press, 1976), p. 3.

3. Edgar Rice Burroughs, *A Princess of Mars* (New York: Ballantine Books, 1963), p. 40.

Chapter 2

1. Phone interview with Clyde Tombaugh, March 31, 1996.
2. Clyde Tombaugh, Ibid.

Chapter 3

1. Wulff Heintz, personal interview, April 17, 1996.
2. Wulff Heintz, Ibid.
3. Wulff Heintz, Ibid.
4. Wulff Heintz, Ibid.
5. Sarah Lee Lippincott, quoted in the *Philadelphia Inquirer*, May 23, 1995, p. B6.

Chapter 4

1. Andrew Lyne. Quoted by Richard Tresch Feinberg in "Pulsars, Planets and Pathos," *Sky and Telescope*, May 1992, p. 494.
2. Andrew Lyne. Quoted by Tim Folger in "Forbidden Planets," *Discover*, Vol. 13, No. 4, April 1992, p. 40.
3. Stan Woosley. Quoted by Tim Folger in "Forbidden Planets," *Discover*, Vol. 13, No. 4, April 1992, p. 41.
4. Alexander Wolszczan, personal communication, June 10, 1995.
5. Alexander Wolszczan, Ibid.
6. Alexander Wolszczan, Ibid.

Chapter 5

1. Geoffrey Marcy, personal communication, October 16, 1996.
2. Geoffrey Marcy, Ibid.
3. Geoffrey Marcy, Ibid.
4. Geoffrey Marcy, Ibid.
5. Geoffrey Marcy, Ibid.

6. Daniel Goldin, speech to the January 1996 meeting of the American Astronomical Society. Reported in the *Philadelphia Inquirer*, January 18, 1996, p. A8.

Chapter 6

1. Fritz Zwicky. Quoted by Barry Parker in *Invisible Matter and the Fate of the Universe* (New York: Plenum, 1989), p. 49.
2. Charles Misner, Kip Thorne, and John Wheeler, *Gravitation* (San Francisco: W. H. Freeman, 1973), p. 5.

Chapter 8

1. Giuseppe Cocconi and Philip Morrison, "Searching for Interstellar Communications," *Nature*, Vol. 184 (1959), p. 844.
2. Frank Drake, "The Search for Extraterrestrial Life," *Los Alamos Science Fellows Colloquium*, Vol. 9 (1988), p. 51.

Conclusion

1. Martyn Fogg, personal communication, January 13, 1997.

RELATED READING

The following is a list of general and technical readings related to the search for new planets and extraterrestrial life. Technical references are indicated with asterisks.

Introduction

Drake, Frank D. and Sobel, Dava, *Is Anyone Out There? The Scientific Search for Extraterrestrial Intelligence* (New York: Delacorte Press, 1992).

Field, George and Goldsmith, Donald, *The Space Telescope: Eyes Above the Atmosphere* (New York: Contemporary Books, 1989).

Chapter 1

Burroughs, Edgar Rice, *A Princess of Mars* (New York: Ballantine Books, 1980).

Hoyt, William Graves, *Lowell and Mars* (Tucson: The University of Arizona Press, 1976).

Kiernan, Vincent, Hecht, Jeff, Cohen, Philip and Concar, David, "Did Martians Land in Antarctica," *New Scientist*, Vol. 151, No. 2043 (1996), pp. 4–5.

Lowell, Percival, *Mars* (London: Longmans, Green and Co., 1895).

Lowell, Percival, *Mars as the Abode of Life* (New York: The Macmillan Company, 1908).

Morrison, David and Owen, Tobias, *The Planetary System* (New York: Addison-Wesley Publishing Company, 1987).

Sheehan, William, *Worlds in the Sky: Planetary Discovery from Earliest Times through Voyager and Magellan* (Tucson: The University of Arizona Press, 1992).

Wells, H. G. "The War of the Worlds," In *Seven Science Fiction Novels of H. G. Wells* (New York: Dover Publications, 1934).

Chapter 2

Holmes, Bob and Schilling, Govert, "Hidden Helium Heats Jupiter from Within," *New Scientist*, Vol. 149, No. 2015 (1996), p. 16.

Hoyt, William Graves, *Planet X and Pluto* (Tucson: The University of Arizona Press, 1980).

Morrison, David and Owen, Tobias, *The Planetary System* (New York: Addison-Wesley Publishing Company, 1987).

Sheehan, William, *Worlds in the Sky: Planetary Discovery from Earliest Times through Voyager and Magellan* (Tucson: The University of Arizona Press, 1992).

Chapter 3

Abt, Herbert, "The Companions of Sunlike Stars," *Scientific American*, Vol. 236 (April 1977), pp. 96–104.

Binder, Otto, *Riddles of Astronomy* (New York: Basic Books, 1964).

*Black, David, *Project Orion: A Design Study for Detecting Extrasolar Planets* (Washington: National Aeronautics and Space Administration, 1980).

Black, David, "Worlds around Other Stars," *Scientific American*, Vol. 250, (January 1991), pp. 76–82.

Bruning, David, "Desperately Seeking Jupiters," *Astronomy*, (July 1992), pp. 37–41.

Dick, Stephen, *Plurality of Worlds* (New York: Cambridge University Press, 1982).

Friedman, Herbert, *The Astronomer's Universe* (New York: Ballantine Books, 1990).

Goldsmith, Donald and Cohen, Nathan, *Mysteries of the Milky Way* (Chicago: Contemporary Books, 1991).

*Heintz, Wulff D., *Double Stars* (Boston: D. Reidel Publishing Co., 1978).

*Heintz, Wulff D., "Reexamination of Suspected Unresolved Binaries," *The Astrophysical Journal*, Vol. 220 (1978), pp. 931–934.

*Heintz, Wulff D., "Systematic Trends in the Motions of Suspected Stellar Companions," *Monthly Notes of the Royal Astronomical Society*, Vol. 175 (1976), pp. 533–535.

Maran, Stephen P. "Stalking the Extrasolar Planet," *Natural History*, Vol. 5, (1989), pp. 70–73.

Motz, Lloyd and Weaver, Jefferson Hane, *The Unfolding Universe: A Stellar Journey* (New York: Plenum, 1989).

Muirden, James, *Stars and Planets* (New York: Thomas Crowell Co., 1964).

Sullivan, Walter, *We Are Not Alone* (New York: Dutton, 1992).

*Tarter, Jill, "The Current State of Searches for Extra-solar Planets," In *Third Decennial US–USSR Conference on SETI, ASP Conference Series*, Vol. 47. Edited by G. Seth Shostak (San Francisco: Astronomical Society of the Pacific, 1993).

*van de Kamp, Peter, "Astrometric Study of Barnard's Star from Plates Taken with the 24-inch Sproul Refractor," *Astronomical Journal*, Vol. 68, No. 7 (1963), pp. 515–522.

van de Kamp, Peter, "Barnard's Star as an Astrometric Binary," *Sky and Telescope* (July 1963), pp. 8–9.

*van de Kamp, Peter, *Dark Companions of Stars* (Boston: D. Reidel Publishing Co., 1986).

Wallace, Alfred Russel, *Man's Place in the Universe* (New York: McClure Phillips and Co., 1904).

Chapter 4

*Bailes, M., Lyne, A.G., and Shemar, S.L. "A Planet Orbiting the Neutron Star PSR 1829-10," *Nature*, Vol. 352 (1991), pp. 311–313.

Bruning, David, "Lost and Found: Pulsar Planets," *Astronomy* (June 1992), pp. 36–38.

Croswell, Ken, "Puzzle of the Pulsar Planets," *New Scientist*, Vol. 137, No. 1832 (1992), pp. 40–43.

Fienberg, Richard Tresch, "Pulsars, Planets and Pathos," *Sky and Telescope*, Vol. 83, (May 1992), pp. 493–495.

Folger, Tim, "Forbidden Planets," *Discover*, Vol. 13 (April 1992), pp. 38–43.

*Gil, J.A. and Jessner, A., "Are There Really Planets around PSR 1257+12." In *Planets around Pulsars*, edited by J.A. Phillips, J.E.

Thorsen and S.R. Kulkarni, ASP Conference Series, Vol. 36 (1993), pp. 71–79.

*Nakamura, Takashi and Piran, Tsvi, "The Origin of the Planet around PSR 1829-10," *The Astrophysical Journal*, Vol. 382 (1991) pp. L81–L84.

*Peale, S.J. "On the Verification of the Planetary System around PSR 1257+12," *The Astronomical Journal*, Vol. 105, No. 4 (1992), pp. 1562–1570.

*Podsiadlowski, P., Pringle, J.E. and Rees, M.J., "The Origin of the Planet Orbiting PSR 1829-10," *Nature*, Vol. 352 (1991), pp. 783–784.

*Tavani, Marco and Brookshaw, Leigh, "The Origin of Planets Orbiting Millisecond Pulsars," *Nature*, Vol. 356 (1992), pp. 320–322.

Travis, John, "Pulsing Star Confirms More Planets in the Universe," *Science*, Vol. 264, No. 16 (1994), pp. 506–507.

Wilford, John Noble, "Growing Proof of Other Worlds Out There," *New York Times* (April 22, 1994), p. 1.

Will, Clifford, "The Good Companions," *Nature*, Vol 355 (1992), pp. 111–113.

*Wolszczan, Alexander, "PSR 1257+12 and Its Planetary Companions." In *Planets around Pulsars*, edited by J. A. Phillips, J. E. Thorsen and S. R. Kulkarni, ASP Conference Series, Vol. 36 (San Francisco: Astronomical Society of the Pacific, 1993), pp. 3–9.

*Wolszczan, Alexander, "Confirmation of Earth-Mass Planets Orbiting the Millisecond Pulsar PSR B1257+12," *Science*, Vol. 264, No. 16 (1994), pp. 538–542.

*Wolszczan, Alexander and Frail, Dale A., "A Planetary System around the Millisecond Pulsar PSR 1257+12," *Nature*, Vol. 355 (1992), pp. 145–149.

Chapter 5

*Angel, J. R. P., "Ground-based Imaging of Extrasolar Planets using Adaptive Optics," *Nature*, Vol. 368 (1994), pp. 203–206.

*Black, David C., "A Comparison of Alternative Methods for Detecting Other Planetary Systems." In *Strategies for the Search for Life in the Universe*, edited by M. D. Papagiannis. (New York: D. Reidel Publishing Co., 1980), pp. 167–175.

Davies, John, "Searching for Alien Earths," *New Scientist*, Vol. 146, No. 1977 (1995), pp. 24–28.

Glanz, James, "Hints of a Planet Orbiting Sunlike Star," *Science*, Vol. 270 (1995), p. 375.

MacRobert, Alan M., and Roth, Joshua, "The Planet of 51 Pegasus, *Sky and Telescope*, Vol. 87 (January 1996), pp. 38–40.

Walker, Gabrielle, "Seven Planets for Seven Stars," *New Scientist*, Vol. 150, No. 2034 (1996), pp. 26–30.

Chapter 6

*Albrow, M., Birch, P., Caldwell, J., Martin, R., Menzies, J., Pel, J.-W., Pollard, K. Sackett, P. D., Sahu, K., Vreeswijk, P., Williams, A., Zwaan, M., "The PLANET Collaboration: Probing Lensing Anomalies with a World-Wide Network," Preprint 213, Kapetyn Institute, Groningen, The Netherlands (1996), pp. 1–2.

*Alcock, C., Bennett, D., Cook, K., Allsman, R., Alves, D., Axelrod, T., Freeman, K., Peterson, B., Rodgers, A., Griest, K., Guern, J., Lehner, M., Quinn P., Marshall, S., Pratt, M., Becker, A., Stubbs, C., and Sutherland, W., "The MACHO Collaboration Search for Baryonic Dark Matter via Gravitational Microlensing," Proceedings of the Pascos/Hopkins Symposium, Baltimore, Maryland, World Scientific (1995).

*Alcock, C., Bennett, D., Cook, K., Allsman, R., Alves, D., Axelrod, T., Freeman, K., Peterson, B., Rodgers, A., Griest, K., Guern, J., Lehner, M., Quinn, P., Marshall, S., Pratt, M., Becker, A., Stubbs, C., and Sutherland, W., "The MACHO Project Second Year Gravitational Microlensing Results Toward the Large Magellanic Cloud," Proceedings of the 187th Meeting of the American Astronomical Society, San Antonio, Texas, (January, 1996).

Bartusiak, Marcia, *Through A Universe Darkly: A Cosmic Tale of Ancient Ethers, Dark Matter and the Fate of the Universe* (New York: Harper-Collins, 1993).

Cowen, R., "Dark Matter: MACHOS in Milky Way's Halo?" *Science News* (September, 1993), p. 199.

Ferris, Timothy, *Coming of Age in the Milky Way* (New York: William Morrow, 1988).

Friedman, Herbert, *The Astronomer's Universe* (New York: Ballantine Books, 1990).

Halpern, Paul, *The Cyclical Serpent: Prospects for an Ever-Repeating Universe* (New York: Plenum, 1995).

The Quest for Alien Planets

Krauss, Lawrence, *The Search for the Fifth Essence: Dark Matter in the Universe* (New York: Basic Books, 1989).

*Mao, Shunde and Paczynski, Bohdan, "Gravitational Microlensing by Double Stars and Planetary Systems," *Astrophysical Journal Letters*, 374 (1991), p. 37.

Parker, Barry, *Invisible Matter and the Fate of the Universe* (New York: Plenum, 1989).

Riordan, Michael and Schramm, David, *The Shadows of Creation: Dark Matter and the Structure of the Universe* (New York: W. H. Freeman, 1991).

Trefil, James, *The Dark Side of the Universe* (New York: Scribners, 1988).

*Trimble, Virginia, "Existence and Nature of Dark Matter in the Universe," In *The Early Universe: Reprints*, edited by Edward Kolb and Michael Turner (New York: Addison-Wesley, 1988).

Tucker, Wallace and Tucker, Karen, *The Dark Matter* (New York: Morrow, 1988).

Tyson, Anthony, "Mapping Dark Matter with Gravitational Lenses," *Physics Today* (June, 1992), pp. 24–25.

*Udalski, A., Szymanski, M., Kaluzny, J., Kubiak, M., and Mateo, M., "The Optical Gravitational Lensing Experiment," *Acta Astronomica*, 42, pp. 253–260.

*Zwicky, Fritz, "Nebulae as Gravitational Lenses," *Physical Review* 51 (February 15, 1937), p. 290.

Chapter 7

*Burke, Bernard F., "Detection of Planetary Systems and the Search for Evidence of Life," *Nature* 322 (July 24, 1986), pp. 340–343.

Dole, Stephen, *Planets for Man* (New York: Random House, 1964).

*Fogg, Martyn J., "An Estimate of the Prevalence of Biocompatible and Habitable Planets," *Journal of the British Interplanetary Society* 45 (1992), pp. 3–12.

*Greenstein, Jesse L., "Search for Planets and Early Life in Other Solar Systems: An Introduction." In *Strategies for the Search for Life in the Universe*, edited by M. D. Papagiannis (New York: D. Reidel Publishing Co., 1980), pp. 109–110.

*Owen, Tobias, "The Search for Early Forms of Life in Other Planetary Systems: Future Possibilities Afforded by Spectroscopic Techniques." In *Strategies for the Search for Life in the Universe*, edited by

M. D. Papagiannis (New York: D. Reidel Publishing Co., 1980), pp. 177–185.

Pollard, William G., "The Prevalence of Earthlike Planets," *American Scientist* 67, (November–December 1979), pp. 653–660.

Roy, A.E. and Clarke, D., *Astronomy: Structure of the Universe* (Bristol: Adam Hilger, Ltd., 1982).

*Schneider, Jean, "Strategies for the Search for Life in the Universe," In *Proceedings of the Conference on Biophysical Evolution of Life, Trieste, September 1995*, edited by J. Chela-Flores and F. Raulin (Boston: Kluwer Academic, 1996), pp. 1–15.

Tutukov, A.V., "Formation of Planetary Systems in the Galaxy," In *Third Decennial US–USSR Conference on SETI*, ASP Conference Series, Vol. 47, edited by G. Seth Shostak (San Francisco: Astronomical Society of the Pacific, 1993), pp. 185–193.

*Wetherill, George W., "Occurrence of Earth-Like Bodies in Planetary Systems," *Science* 253 (August 2, 1991), pp. 535–538.

Chapter 8

*Blair, D. G. and Zadnik, M. G., "A List of Possible Interstellar Communication Channel Frequencies for SETI," *Astronomy and Astrophysics* 278 (1993), pp. 669–672.

Brin, David, "Mystery of the Great Silence," In *First Contact: The Search for Extraterrestrial Intelligence*, edited by Ben Bova and Byron Preiss. (New York: Plume, 1990), pp. 118–140.

*Cocconi, Giuseppe and Morrison, Philip, "Searching for Interstellar Communications, *Nature* 184 (1959), p. 844.

Drake, Frank, "A Brief History of SETI," In *Third Decennial US-USSR Conference on SETI*, ASP Conference Series, Vol. 47, edited by G. Seth Shostak (San Francisco: Astronomical Society of the Pacific, 1993), pp. 11–17.

*Gindilis, I. M., Davyov, V. P., and Strelnitski, V. S., "New 'Magic' Frequencies for SETI," In *Third Decennial US-USSR Conference on SETI*, ASP Conference Series, Vol. 47, edited by G. Seth Shostak, (San Francisco: Astronomical Society of the Pacific, 1993), pp. 161–163.

Gribbin, John, "Is Anyone Out There?" *New Scientist*, Vol. 145 (25 May, 1991), pp. 26–30.

Naeye, Robert, "SETI at the Crossroads," *Sky and Telescope* (November, 1992), pp. 507–515.

*Oliver, B. M., "Rationale for the Water Hole," *Acta Astronautica* 6 (1979), pp. 71–79.

Sagan, Carl, *Cosmos*, (New York: Random House, 1980).

Shklovskii, I. S. and Sagan, Carl, *Intelligent Life in the Universe* (San Francisco: Holden Davy, Inc., 1966).

Tipler, Frank, "Extraterrestrial Intelligent Beings Do Not Exist," *Quarterly Journal of the Royal Astronomical Society* 21 (1981), pp. 267–282.

White, Frank, *The SETI Factor* (New York: Walker and Co., 1990).

Conclusion

Halpern, Paul, *Cosmic Wormholes: The Search for Interstellar Shortcuts* (New York: Dutton, 1992).

PLANET-RELATED
COMPUTER SITES ON
THE WORLD WIDE WEB

Here is a select group of World Wide Web sites pertaining to the search for extrasolar planets. Note that the World Wide Web is a dynamic network in an almost constant state of flux; therefore I cannot guarantee that each of these sites still operates. Furthermore, because the Web is not professionally refereed, I cannot vouch for the accuracy of the information contained in these sources. I provide these listings merely to guide the reader toward up-to-date information about a rapidly expanding field of knowledge.

DARWIN PROJECT:
http://ast.star.rl.ac.uk/darwin/

EXPLORATION OF NEIGHBORING PLANETARY SYSTEMS:
http://techinfo.jpl.nasa.gov/WWW/ExNPS/
RoadMap.html

EXTRASOLAR PLANETS ENCYCLOPEDIA:
http://www.obspm.fr/departement/darc/planets/encycl.html

KEPLER MISSION:
http://www.kepler.arc.nasa.gov

PULSAR PLANETS PAGE:
http://www.astro.psu.edu/users/pspm/arecibo/planets/planets.html

SAN FRANCISCO STATE UNIVERSITY PLANET SEARCH PROJECT:
http://cannon.sfsu.edu/~williams/planetsearch/planetsearch.html

SETI INSTITUTE:
http://www.seti-inst.edu

INDEX